T0176441

LECTURE NOTES ON IMPEDANCE SPECTROSCOPY

Lecture Notes on Impedance Spectroscopy

Measurement, Modeling and Applications

Editor

Olfa Kanoun

Chair for Measurement and Sensor Technology
Chemnitz University of Technology, Chemnitz, Germany

VOLUME 3

CRC Press
Taylor & Francis Group
Boca Raton   London   New York   Leiden

CRC Press is an imprint of the
Taylor & Francis Group, an **informa** business

A BALKEMA BOOK

CRC Press/Balkema is an imprint of the Taylor & Francis Group, an informa business

© 2012 Taylor & Francis Group, London, UK

Typeset by V Publishing Solutions Pvt Ltd., Chennai, India
Printed and bound in Great Britain by CPI Group (UK) Ltd, Croydon, CR0 4YY

All rights reserved. No part of this publication or the information contained herein may be reproduced, stored in a retrieval system, or transmitted in any form or by any means, electronic, mechanical, by photocopying, recording or otherwise, without written prior permission from the publisher.

Although all care is taken to ensure integrity and the quality of this publication and the information herein, no responsibility is assumed by the publishers nor the author for any damage to the property or persons as a result of operation or use of this publication and/or the information contained herein.

Published by: CRC Press/Balkema
            P.O. Box 447, 2300 AK Leiden, The Netherlands
            e-mail: Pub.NL@taylorandfrancis.com
            www.crcpress.com – www.taylorandfrancis.com

ISBN: 978-0-415-64430-3 (Hbk)
ISBN: 978-0-203-07512-8 (eBook)

Lecture Notes on Impedance Spectroscopy, Volume 3 – Kanoun (ed)
© 2012 Taylor & Francis Group, London, ISBN 978-0-415-64430-3

Table of contents

Lecture Notes on Impedance Spectroscopy, Volume 3 – Kanoun (ed)
© 2012 Taylor & Francis Group, London, ISBN 978-0-415-64430-3

Preface

Impedance Spectroscopy is a powerful measurement method used in many application fields such as electrochemistry, material science, biology and medicine. The possibility to use information from complex impedance over a wide frequency range leads to interesting opportunities for separating effects, accurate measurements and measurements of non-accessible quantities.

Using Electrochemical Impedance Spectroscopy (EIS) in general requires competences in several fields of science and technology. It includes developing a suitable measurement procedure, understanding the electro-chemical and physical phenomena taking place, developing suitable models and extracting the target information from models or by using sophisticated mathematical methods. Depending on the specific challenges of the considered device under test there are generally more efforts to be done in one or two specific fields. The scientific dialogue between specialists in Impedance Spectroscopy, dealing with different application fields, is therefore particularly important to promote the adequate use of this powerful measurement method in both laboratory and in embedded solutions.

The international Workshop on Impedance spectroscopy (IWIS) has established itself as a platform for promoting experience exchange and networking in the scientific and industrial field. It has been launched in 2008 with the aim to serve as a platform for specialists and users to share experiences with each other. The workshop has been gaining increasingly more acceptance in the scientific and industrial fields and addressing increasingly more fundamentals and diverse application fields of impedance spectroscopy.

This book is the third in the series Lecture Notes on Impedance Spectroscopy. It includes selected and extended contributions from the International Workshop on Impedance Spectroscopy (IWIS'11). It is a set of presented contributions of world-class manuscripts describing state-of-the-art research in the field of impedance spectroscopy. It reports about new advances and different approaches in dealing with impedance spectroscopy including theory, methods and applications. The book is interesting for researchers and developers in the field of impedance spectroscopy.

I thank all contributors for the interesting contributions and for their confidence in us during the preparation of the proceedings.

Prof. Dr.-Ing. Olfa Kanoun

Lecture Notes on Impedance Spectroscopy, Volume 3 – Kanoun (ed)
© *2012 Taylor & Francis Group, London, ISBN 978-0-415-64430-3*

Distance measure for impedance spectra for quantified evaluations

Meike Slocinski, Kathrina Kögel & Johann-Friedrich Luy
Daimler AG, Germany

ABSTRACT: Electrochemical impedance spectroscopy is a measurement method for detailed investigations of electrochemical systems. The increasing number of recorded impedance spectra requires automated evaluations. Manual evaluations based on the visual comparisons of several impedance spectra in a Nyquist plot are subjective, time-consuming and thus no longer feasible. The challenge is to find quantitative, reproducible and therefore objective evaluation methods that coincide with visual impressions.

In this work a distance measure for the space of impedance spectra is developed which formalizes the concept of similarity. Its origin is the mathematical description of an impedance spectrum as a function. Interactions between the distribution of the measurement frequencies and weighting factors are clarified. Furthermore, the importance of a distance measure for various applications is discussed.

Keywords: impedance spectrum, distance measure, metric, norm, weighted distance, logarithmic frequency scale.

1 INTRODUCTION

Electrochemical impedance spectroscopy (EIS) is an established method for the investigation of a wide range of electrochemical systems (Barsoukov and Macdonald 2005). A prerequisite for many evaluation methods is the analysis of the similarity of two impedance spectra for example when assessing the quality of fit or when describing the magnitude of state changes. This requires a scalar measure for the distance between two impedance spectra.

A common application for EIS is the investigation of Li-ion cells and batteries. E-mobility will play a key role in the automotive industry in the next decades. This involves the usage of a large number of cells and batteries for traction. If impedance spectroscopy qualifies as a standard measurement technique for the characterization of these energy storage components, a large number of impedance spectra will be generated. If the chemical aspects behind a spectrum are of minor interest, it is sufficient to treat the impedance spectra simply as a numerical source of information. In contrast to the laboratory environment, manual analysis of the impedance spectra by experts is no longer feasible, and thus objective quantitative methods must be developed to evaluate measured impedance spectra quickly and automated allowing for large-scale data analysis.

In this work, functional distance measures for the complex impedance are suggested based on the definitions of norm and metric. The discretized version of the distance measure is derived by using numerical integration in the calculation step of the functional distance measure.

Impedance spectra are measured on a discrete grid of frequencies yielding vectors of measured impedance points. For the discrete vector distance, however the choice of measurement frequencies is of particular importance. Implicit weighting effects arise which are analyzed and explicit weighting factors are introduced. Furthermore, possible applications which require distance measures are proposed.

2 MATHEMATICAL PREREQUISITS

2.1 *Mathematical analysis of impedance spectra*

The impedance $Z(\omega)$ of a system is a continuous complex function on the positive frequency domain, see Equation (2). This continuous function is sampled on a number of angular frequencies $\omega_i = 2\pi f_i$ by measuring the corresponding impedance points $Z(\omega_i)$ in the complex plane. Hence, an impedance point consists of a real part $Z^{Re}(\omega_i)$ and an imaginary part $Z^{Im}(\omega_i)$ and is thus 2-dimensional. The index i denotes the ith frequency of the discretized impedance spectrum.

$$Z : \mathbb{R}^+ \mapsto \mathbb{C}$$
$$\omega \mapsto \begin{pmatrix} Z^{Re}(\omega) \\ Z^{Im}(\omega) \end{pmatrix} \tag{1}$$

The impedance curve measured is independent from number and distribution of frequencies in the support vector. Discretization of an impedance spectrum originates only due to the measurement process or when a functional model equation for the impedance is resolved on an arbitrary dense grid of frequencies. By using smoothing or interpolation methods, e.g. roughness penalty method (Green and Silverman 2000), missing impedance points or the entire curve can be estimated from such a discretized version, i.e. a vector of impedance points.

2.2 *Requirements for a distance measure*

Since impedance spectra are continuous functions, determining the distance between two impedance spectra means determining the distance between functions. A reasonable definition of a distance is a metric. Any metric on the function space of impedance spectra must fulfill the following three axioms (A1)–(A3) for the distance d for arbitrary continuous functions f, g and h:

- Positive definiteness—a distance is always positive-valued and equal to zero if and only if f and g are identical:

$$d(f,g) \geq 0 \quad \text{and} \quad d(f,g) = 0 \iff f = g \tag{A1}$$

- Symmetry—interchanging functions f and g does not affect the value of the distance:

$$d(f, g) = d(g, f) \tag{A2}$$

- Triangle inequality—the distance between two functions f and g is always less than or equal to the distance of a detour via a third function h:

$$d(f, g) \leq d(f, h) + d(h, g) \tag{A3}$$

These are general requirements for a distance measure that meets the human intuition for distances, for example when observing impedance spectra in Nyquist plots.

A metric can always be defined as the norm of the difference of two functions f and g, see Equation (2).

$$d(f,g) = \|f - g\| \tag{2}$$

For functions, a distance $\delta(s)$ between the values $f(s)$ and $g(s)$ for any s is taken. This requires a metric on the space of function values, called inner distance. On the function $\delta(s)$ which takes values in the positive real numbers, a functional norm, the outer norm, is applied.

2

Table 1. Commonly used norms in functional and discrete form.

Based on	Functional	Discrete
1-norm	$\int_{S_1}^{S_2}\|f(s)-g(s)\|ds$	$\sum_{i=1}^{N}\|f_i-g_i\|$
2-norm	$\sqrt{\int_{S_1}^{S_2}(f(s)-g(s))^2 ds}$	$\sqrt{\sum_{i=1}^{N}(f_i-g_i)^2}$
Infinity norm	$\sup_s\{(f(s)-g(s))\}$	$\sup_i\{(f_i-g_i)\}$

2.3 Norms

Distances can be calculated based on different norms. Equation (7) shows common functional norms of order p for a function f.

$$\|f\|_p=\left(\int f^p\right)^{\frac{1}{p}} \tag{3}$$

The most commonly used norms for constructing metrics are the special cases 1-norm ($p=1$), 2-norm ($p=2$) and infinity norm. The formulas for the corresponding functional distance measures are listed in the first column of Table 1. In the second column the equivalent discrete distance measures for vectors $(f_1,...,f_N)$ and $(g_1,...,g_N)$ are given.

For given continuous functions f and g the integrals are evaluated by numerical integration methods. These methods always involve the resolution of the functions on a dense grid.

The functional distance measures considered are defined for functions on the same domain, the interval $[S_1, S_2]$. If impedance spectra are measured on different domains, the distance can only be calculated for the overlapping area of the frequency interval. The distribution of the frequencies within the overlapping interval is of minor interest. If the spectra are measured on the same frequency interval but on different frequency grids they can be transformed to functional form, as mentioned above, by using smoothing or interpolation methods. The bandwidth is not a parameter of the distance measure, it is rather a fixed value which has to be stated together with the distance measure applied. In the following it is assumed that all impedance spectra are measured on the same bandwidth.

3 DISTANCE MEASURE FOR IMPEDANCE SPECTRA

3.1 Functional distance measure

For the definition of a distance measure for two impedance spectra Z_1 and Z_2, the complex nature of an impedance point $Z(\omega)$ has to be taken into account. The distance between the two complex impedance points at a certain frequency ω, the inner distance $\delta(\omega_i)$, is determined first. The complex impedance points are two-dimensional vectors containing real and imaginary part of the impedance, see the second column of Table 1 for examples of vector distances. This inner distance evaluated at all frequencies is then a real-valued function over the frequency domain on which then an outer norm is applied. This outer norm includes an integral over the inner distance function $\delta(\omega)$.

For the definition of the distance measure any combination of inner distance on the complex plane and outer norm on the real-valued function can be utilized (see Section 2.3). Table 3.1 shows the functional distance measures for combinations of 1-norm and 2-norm for inner distance and outer norm, respectively. The first index of d denotes the order of the outer norm and the second index the order of the norm on which the inner distance is based.

3

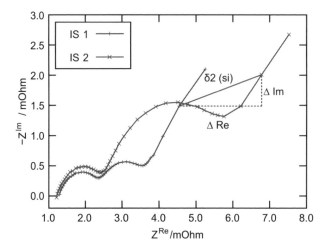

Figure 1. The inner distance $\delta_2(s_i)$, which is the Euclidean distance in the complex plane, between two impedance spectra at a certain frequency s_i.

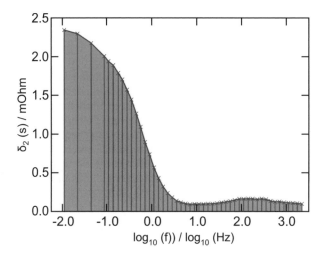

Figure 2. Outer 1-norm over the inner distance function $\delta_2(s)$, which is the integral over inner distance.

Although any metric can be utilized for the inner distance on the complex plane, the Euclidean distance based on the 2-norm is usually taken as the distance between two complex numbers, see Equation (8) and Figure 1. The variable s denotes a frequency scale in general and stands for all possible frequency scales, see Section 3.4.

$$\delta_2(s) = \sqrt{\left(f^{Re}(s) - g^{Re}(s)\right)^2 + \left(f^{Im}(s) - g^{Im}(s)\right)^2} \tag{4}$$

Taking the 1-norm as the outer norm, which is a simple integration over the inner distance, leads to the functional distance measure $d_{1,2}$ in Equation (9) and Figure 2, which is used for the following investigations.

$$d_{1,2}(f,g) = \left\| \delta_2(s) \right\|_1 = \int_{S_1}^{S_2} \sqrt{\left(f^{Re}(s) - g^{Re}(s)\right)^2 + \left(f^{Im}(s) - g^{Im}(s)\right)^2}\, ds \tag{5}$$

Figure 3 shows the illustration of this distance measure in the Nyquist plot.

Table 2. Possible combinations for functional distance measures of inner distance and outer norm for 1-norm and 2-norm, respectively. The first index of d denotes the order of the outer norm over the inner distances and the second index the order of the norm of the inner distance. With $\Delta Z^{RE}(s) = f^{RE}(s) - g^{RE}(s)$ and $\Delta Z^{IM}(s) = f^{IM}(s) - g^{IM}(s)$.

	Inner distance based on 1-norm	Inner distance based on 2-norm
Outer norm 1-norm	$d_{1,1}(f,g) = \int_{S_1}^{S_2} \left\| \left\| \Delta Z^{Re}(s) \right\| + \left\| \Delta Z^{Im}(s) \right\| \right\| ds$	$d_{1,2}(f,g) = \int_{S_1}^{S_2} \sqrt{(\Delta Z^{Re}(s))^2 + (\Delta Z^{Im}(s))^2} \, ds$
Outer norm 2-norm	$d_{2,1}(f,g) = \sqrt{\int_{S_1}^{S_2} \left[|\Delta Z^{Re}(s)| + |\Delta Z^{Im}(s)| \right]^2 ds}$	$d_{2,2}(f,g) = \sqrt{\int_{S_1}^{S_2} (\Delta Z^{Re}(s))^2 + (\Delta Z^{Im}(s))^2 \, ds}$

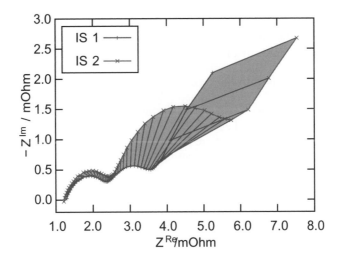

Figure 3. Graphical representation of functional distance measure between two impedance spectra in a Nyquist plot. The bold connections represent examples of inner distances between impedance points at the same frequency.

3.2 Discretized distance measure

The discretized distance measure is derived by evaluation of the integral in the functional distance measure. This is accomplished by applying numerical integration methods, e.g. rectangle rule, trapezoidal rule or Simpson rule. Using the rectangle rule and taking the left-hand sum of the functional distance measure in Equation (5) leads to the discretized distance measure, see Equation (6).

$$d_{1,2}(f,g) \approx \sum_{i=1}^{N} \delta_2(s_i) \underbrace{|s_i - s_{i-1}|}_{:=\Delta_i s} \tag{6}$$

where $i = 1 \ldots N$ and N is the number of measurement points and $\Delta_i s$ the interval width left of the ith measurement point.

It can be observed from this equation that, compared to the discrete vector distance, the Euclidean distances $\delta_2(s_i)$ are weighted with the widths of the intervals $\Delta_i s$. In case of equidistant measurement frequencies the interval widths and thus the weighting is constant. In case of non-equidistant measurement frequencies the weights vary depending on the interval widths and thus on the distribution of the measurement points. In the limiting case, for infinitesimal small interval widths, the discretized distance measure approaches the functional distance measure.

5

Since the recorded impedance spectra are sampled representatives of the functional impedance, the discretized version of the distance measure can be applied directly for calculating distances between two measured impedance spectra.

3.3 *Weighting and weighting factor*

Weighting means multiplying the inner distances $\delta(s)$ with a certain frequency-depending weighting factor $w(s)$. Weighting is always included and can occur due to various effects. Uniform weighting means, that the impedance at any frequency is weighted with constant weighting factor $w(s)$, w.l.o.g. $w(s) \equiv 1$.

In the following it is distinguished between implicit and explicit weighting. If the measurement frequencies are equidistantly distributed in the discretized distance measure, i.e. constant interval widths, uniform weighting is equivalent to simply summing up over discrete frequencies up to a constant factor, i.e. applying the discrete vector metric on the vectors of measured impedance points. This is a kind of incomplete numerical integration, neglecting the interval widths. If the frequencies are not equidistantly distributed on the measurement scale and the discrete vector metric is applied, areas with high measurement frequency density have more influence on the value of the distance. This can be seen as an implicit or intrinsic weighting, depending on the distribution of the measurement frequencies. If the distribution is non-continuous over the frequency domain, the weighting over the frequency will be non-continuous and there is often no evident reason why a certain frequency interval is weighted higher than the frequency interval next to it.

If weighting is desired an explicit frequency depending weighting factor $w(s)$ can be included, which facilitates continuous weighting.

3.4 *Working on transformed frequency scales*

A measurement point vector contains N measured impedance points, that can be displayed on different frequency scales. Figure 4 illustrates a real part of an impedance spectrum displayed over an exemplary reference scale a with equidistantly distributed frequencies in units of [log(Hz)] and a frequency scale b containing the corresponding frequency values in units of [Hz]. The values of the complex impedance points, and thus the inner distances, are the same for the ith frequency on both scales.

The main difference between the functional distance and the discrete vector distance is the consideration of the distribution of measurement frequencies. It has effect on the value of the distance when evaluating the functional distance measure on transformed scales. A decision must be made which scale is the reference scale a to work on. The weight on the reference scale is defined to be uniform, i.e. $w_a(a) \equiv 1$. Not necessarily but often the measurement frequencies are equidistantly distributed on this scale a.

The functional relationship U for the transformation of the frequencies on scale a to frequencies on scale b is expressed in Equation (7) and (8). With U being a strictly increasing function.

$$a = U(b) \tag{7}$$

$$b = U^{-1}(a) \tag{8}$$

In the following $f_a(a) = f_a(U(b)) = f_b(b)$, i.e. the function values are the same regardless of which scale they are displayed on, i.e. $f_b = f_a \circ U$.

The distance measure calculated on the reference scale a is the reference distance. To achieve the same value for the distance when working with transformed frequencies on scale b, the functional distance measure on scale b has to be weighted by an explicit weighting factor $w_b(b)$. The functional distance measure on scale a with uniform weights is transformed by substitutions in Equation (9).

$$d_{1,2}(f_a, g_a) = \int_{A_1}^{A_2} |f_a(a) - g_a(a)| \, da \tag{9}$$

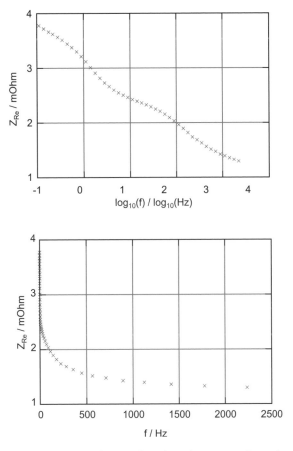

Figure 4. The upper plot shows a real part of an impedance over the scale a with equidistantly distributed logarithmic frequencies. The lower plot shows the corresponding function on the non-logarithmic scale b. The function values are the same for both plots, only the scales are transformed using the relationship $a = \log_{10}(b)$.

The interval limits $[A_1, A_2]$, as well as the impedances at the ith frequency $f_a(a)$ and $g_a(a)$, which are in fact the same as $f_b(b)$ and $g_b(b)$ as mentioned above, and the differential of the integral da are substituted using the functional relationship U between scale a and scale b. This leads to the following conversions:

$$A_1 \rightarrow B_1 = U^{-1}(A_1)$$
$$A_2 \rightarrow B_2 = U^{-1}(A_1)$$
$$f_a(a) \rightarrow f_b(b) = f_a(a)$$
$$g_a(a) \rightarrow g_b(b) = g_a(a)$$
$$da \rightarrow w_b(b)db = d(U(b)) = U'(b)db$$

Equation (10) shows the corresponding distance measure for working on scale b.

$$d_{1,2}(f_b, g_b) = \int_{B_1}^{B_2} |f_b(b) - g_b(b)| w_b(b) db \tag{10}$$

From the substitution of the differential da the weighting factor $w_b(b)$ is obtained. This weighting removes the implicit weighting that occurs when working on a frequency scale other than the reference scale. Working on the scale b with this weighting factor leads to the same distance measure as working on the scale a with uniform weight. This has to be regarded, if the measurement frequencies are not given in terms of the reference scale. For that case either the frequencies have to be converted to frequencies values on the reference scale or the distance measure for working on scale b, including the weighting factor $w_b(b)$, has to be utilized.

If the weighting factor $w_b(b)$ is neglected when working on scale b, which means working on scale b with uniform weights, the distance measure is dominated by effects occurring at areas where the slope of the U-function is small, while effects occurring at areas where the slope of the U-function is steep are nearly not noticeable.

3.5 Example: logarithmic and non-logarithmic scale

In case of investigations made on Li-ion cells the chosen reference scale a is often the logarithmic frequency scale containing frequencies of unit [log(Hz)]. Choosing a frequency vector on this scale with equidistantly distributed logarithmic frequencies leads to measurement points approximately equidistantly distributed on the trajectory of the impedance spectra.

The functional relationship between the reference scale a and the corresponding scale b, that contains frequency values in units of [Hz], is described by Equation (11) and (12).

$$a = \log_{10}(b) \tag{11}$$

$$b = 10^a \tag{12}$$

The substitution of the functional distance measure on the logarithmic scale a in Equation (9), leads to the functional distance measure on the non-logarithmic scale b, see Equation (13).

$$d_{1,2}(f_b, g_b) = \int_{B_1}^{B_2} |f_b(b) - g_b(b)| d(\log_{10}(b))$$
$$= \int_{B_1}^{B_2} |f_b(b) - g_b(b)| \underbrace{\frac{1}{b \cdot \ln(10)}}_{:=w_b(b)} db \tag{13}$$

From the substitution of the differential the weighting factor $w_b(b) = \frac{1}{b \cdot \ln(10)}$ occurs. This weighting factor must be used for working on the non-logarithmic scale b, when the reference scale for uniform weighting is the logarithmic scale a. The use of this weighting factor leads to a distance, which is the same as the reference distance defined on the logarithmic reference scale a.

If the weighting factor $w_b(b)$ is neglected when working on the non-logarithmic scale b, which means working on scale b with uniform weights, the distance measure is dominated by effects occurring at high frequencies, while effects at low frequencies are nearly not noticeable.

4   APPLICATIONS

Distance measures are needed for a wide range of applications, whenever the similarity of two impedance spectra is of interest and the output shall be in form of a scalar value.

The following section gives examples of use cases that require a distance measure: clustering and classification of impedance spectra, investigation of systematic changes of a system, modeling of impedance spectra and the comparison of impedance spectra on different frequency grids.

4.1   Classification and clustering

If impedance spectra are recorded in an automotive context, a large amount of data is expected representing cells or batteries of all kinds of qualities and states. Impedance spectra from cells with similar history are expected to be similar. Accordingly, it is of interest to investigate subgroups of similar impedance spectra. Such subgroups can be defined by using clustering algorithms. They find groups of impedance spectra where the distance between members of the same group is small and the distance between members of different groups is large. This analysis can surely not be accomplished by manual evaluation, rather this automated evaluation requires an objective measure, e.g. the proposed distance measure.

4.2   *Investigation of systematic changes*

In the laboratory environment it is often the task to monitor how systematic changes of an adjustable influence are influencing the current state of a cell. Adjustable influences are e.g. temperature, state of charge or age. For that purpose the cell is exposed to an influence, while keeping other influences constant. Its reaction is then recorded by means of impedance spectroscopy. Figure 5 shows impedance spectra from a cell at five different temperatures, but constant state of charge and age. It can be observed that the impedance spectra change systematically due to the changes of the adjustable influence. The comparison of impedance spectra corresponding to certain states of the Li-ion cell using distances means quantifying its state changes. The distances are listed in a distance table, see Table 3.

Distance tables illustrate the mathematical properties of the distance measure: distance values are positive and zero only if the same impedance spectra are compared, the matrix is symmetric and the triangular inequality is valid for all distances. In this particular case it can be observed that not only the triangular inequality holds, but the distance from 10°C

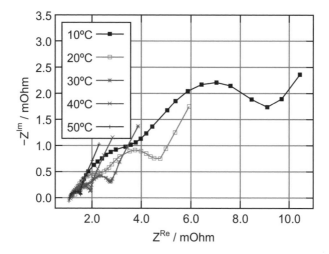

Figure 5.   Impedance spectra from a Li-ion cell at five different temperatures, but constant state of charge and age.

Table 3.   This distance table shows the distances, calculated with the proposed distance measure $d_{1,2}$ from Equation (5), from the impedance spectra shown in Figure 5 due to variation of temperature as an adjustable influence. The values are given in $[10^{-3}]$.

	10°C	20°C	30°C	40°C	50°C
10°C	0	8.31	12.73	15.19	16.57
20°C	8.31	0	4.44	6.92	8.32
30°C	12.73	4.44	0	2.49	3.90
40°C	15.19	6.92	2.49	0	1.41
50°C	16.57	8.32	3.90	1.41	0

to 50°C is approximately the sum of the distance between 10°C and 30°C as well as 30°C and 50°C, which is a nice intuitive property since 30°C is obviously a direct intermediate state between 10°C and 50°C.

4.3 Modeling

To extract the information contained in an impedance spectrum it is modeled, e.g. with an equivalent electrical circuit model (Roth, Slocinski and Luy 2010). This equivalent electrical circuit model provides a model equation with parameters to be adjusted with the goal to minimize the distance between fit and measurement. The fitting procedure requires an optimization algorithm which itself requires a cost function that provides information about the quality of the current fit during the adjustment of parameters to the measurement. The modeling procedure is illustrated in Figure 6.

In order to fit all measured frequencies with the desired weighting, it is recommended to use the proposed distance measure as cost function. Otherwise when simply summing up over the distances at the measurement frequencies an intrinsic weighting is included in the modeling result leading to areas in the impedance spectrum better fitted than others, see Section 3.3.

The output of the modeling procedure is a set of adjusted parameters, describing together with the model's equation the measured impedance spectrum. Additionally, the associated error, that is the distance between fit and measurement is provided, which is a scalar measure for the accuracy of the fitting result.

4.4 Arbitrary frequency grids

The purpose of recording impedance spectra is to collect information about the complex impedance of a system. The information of an impedance spectrum is contained in its curvature. For that reason it might be useful to measure more impedance points in areas of strong curvature. This leads to non-equidistant frequency vectors.

If the distance between two impedance spectra measured on different frequency vectors is to be calculated, the impedance spectra need to be preprocessed. Therefore the impedance spectra are to be functional approximated, e.g. by utilizing smoothing and interpolation methods, e.g. roughness penalty approach. After that, the impedance spectra can be resolved on equal grids and the functional distance measure can be calculated over an mutual bandwidth.

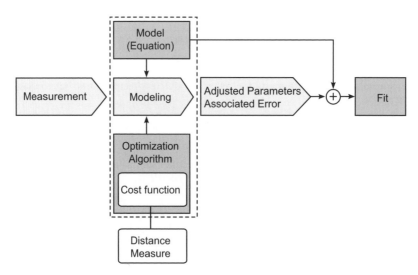

Figure 6. Modeling procedure for the fitting of impedance spectra. The optimization algorithm requires a cost function, therefore the distance measure can be applied.

This case might also occur if either the measurement setup switches frequency vectors, or if impedance spectra from different research groups are to be compared, since they usually do not use the same frequency vectors. In these cases a functional view onthe measurement, as given in this work, is indispensable.

5   CONCLUSION

In this work a distance measure was developed based on the functional description of an impedance spectrum and by deciding upon an inner distance and an outer norm. From the functional distance measure the discretized distance measure was derived by numerical integration. It was shown that such a distance measure is required for a wide spectrum of applications, whenever the similarity of two impedance spectra is of interest and an objective scalar measure is required for the evaluations.

Furthermore it was shown that a distance measure, which is calculated by summing up over the inner distances regardless of the distribution of measurement frequencies, is a weighted functional distance measure. Intrinsic weighting also occurs when workingon a transformed frequency scale, which is not the scale on which it is defined to have uniform weighting.

REFERENCES

E. Barsoukov and J.R. Macdonald, *Impedance Spectroscopy*. John Wiley & Sons, 2005.

P.J. Green and B.W. Silverman, *Nonparametric Regression and Generalized Linear Models*, 1st ed., ser. Monographs on Statistics and Applied Probability 58. Chapman & Hall, 2000.

J.G. Roth, M. Slocinski, and J.-F. Luy, "Modeling li-ion battery aging data utilizing particle swarm optimization," in *Lecture Notes on Impedance Spectroscopy*, O. Kanoun, Ed. CRC Press, 2010, pp. 1–11.

Lecture Notes on Impedance Spectroscopy, Volume 3 – Kanoun (ed)
© 2012 Taylor & Francis Group, London, ISBN 978-0-415-64430-3

Material characterisation of nano scale solid state electrolytes

Frank Wendler, Markus Freund & Olfa Kanoun
Chair for Measurement and Sensor Technology, Chemnitz University of Technology, Chemnitz, Germany

Jörg Schadewald & Carlos Cesar Bof Bufon
Institute for Integrative Nanosciences, IFW Dresden, Dresden, Germany

Oliver G. Schmidt
Institute for Integrative Nanosciences, IFW Dresden, Dresden, Germany
Chair for Material Systems for Nanoelectronics, Chemnitz University of Technology, Chemnitz, Germany

ABSTRACT: Impedance spectroscopy is widely used in material characterization and characterization of energy storage devices. In material analysis the specific properties of macro scale samples are in focus of investigation. However, the characteristic properties obtained with bulk measurements can differ from those measured at thin films. In the field of energy storage devices impedance spectroscopy is used to find parameter sets for modeling the individual storage behavior. Here the dimension of the device are of little interest. The combination of material as well as storage behavior characterization can be used in the case of chip level energy storage devices. For this, test structures are measured over a wide frequency range and analyzed via model based approach. With this process dielectric, resistive and electrochemical effects can be separated and analyzed in a quantitative way. The procedure was tested on nanometer range thin films of the solid state electrolyte LiPON based capacitors. Precise measurement of the energy storage material characteristics can support further optimization of the storage material and production processes.

Keywords: material charaterization, impedance spectroscopy, limited diffusion, solidstate electrolyte

1 INTRODUCTION

Spectral methods are commonly used in the characterization and analysis of unknown systems. For linear systems the frequency dependend relation between output and input, called transfer function is used. For an electrical system this transfer function is the complex impedance. The electrical impedance is close related the power absorption of the system. Therefore the impedance is a suitable property for the characterization of systems for energy storage and conversion. Additionally, the different effects causing power absorbtion and hence a change in impedance are present in different frequency ranges. Thus it is possible to distinguish them with suitable methods if the system is measured over a wide frequency range. For non-overlapping effects simple methods like corner-frequency analysis, graphical analysis and others exist. For overlapping effects, which is most often the case other methods must be used. In this work electrochemical models are used to describe the impedance behavior over a wide frequency range. With parameter extraction methods a set of parameters is extracted with the prior knowledge of the model. They includes the useful information of the spectra. Different types of models exist. Some use equivalent circuits that reduce electrochemical effects into circuit elements. Sometimes systems of differential equations are used. Some models involve more abstract concepts like Voigts circuits, that are used to explain the distribution of relaxation times concept. The first two model typed directly contain an analysis of the sample's impedance shape and structure and with this a representation of one or more processes of power storage and consumption. The chosen model defines also the analytic background for the interpretation of the extracted parameters and can provide implicit predictions of the model, like charge distributions in some cases.

2 EXPERIMENTAL SETUP

The samples consist of a 20 nm lithium phosphorus oxynitride (LiPON) layer incorporated between bottom and top electrodes. The LiPON film was deposited by radio-frequency magnetron sputtering from a Li_3PO_4 target (Kurt J. Lesker company) under a nitrogen flow of 10 sc/cm. Bottom and top electrodes were prepared by direct-current (DC) magnetron sputtering of chromium with a thickness of 30 nm. Contact pads with a thickness of 50 nm were created by DC sputtering of aluminium. The electrode couple covered an area of 200×300 m, illustrated in Figure 2 and Figure 3. Each sample was connected via controlled impedance ceramic blade probes to the Agilent 4294 impedance analyzer, using the 42941 A Impedance Probe Kit, outlined in Figure 1. The total frequency range of data acquisition was 40 Hz to 110 MHz potentiostatic mode with 50 mV peak to peak excitation amplitude.

3 PARAMETER EXTRACTION

The stochastic optimization algorithm particle filter, (Büschel, Tröltzsch, and Kanoun 2011), was used to fit the model to the individual data sets.

4 EFFECTS AND MODELS

The Randles Model is one of the widely used impedance models in the field of electrochemical energy storages devices. The model combines effects of charge transfer at the electrode-electrolyte interface, represented by a charge transfer resistance and a double layer capacitance formed at the elctrode surface, together with effects of ionic diffusion in the diffuse double layer represented by a Warburg impedance, (Hamann and Vielstich 2005). At low frequencies the warburg impedance predicts a characteristic 45°-slope of the impedance in the complex impedance plane. At high frequencies ionic transport and agglomeration processes are expected to be greatly reduced and will result in the characteristic semi circle of a regular capacitance. A modified Randles Circuit is applied as model function in Figure 4.

The model is extended by a constant phase element (CPE), to include the possible effect of inhomogeneity of time constants (Bisquert and Compte 2001). The coefficient β is used

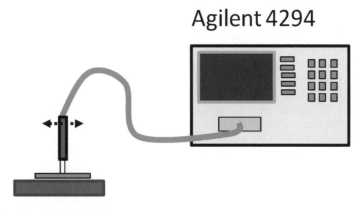

Figure 1.   Outline of the arrangement.

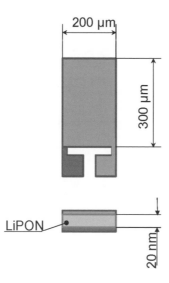

Figure 2.   Size and shape of the sample.

Figure 3.   The sample connected by probe tips.

like in (Tröltzsch 2005) to adjust the angle of the Warburg part to other than 45° in order to create an additional degree of freedom for the fitting algorithm. This model provides a good match for the high and medium frequency range. The additional factors are close to the ideal Randles Model's values of $\beta = 1$ and $\alpha = 0{,}5$ so that obtained parameters can be discussed according to this model. The error in the low frequency range is regarded as additional effect and is resulting in a second part of the graph, with a slope bigger than 45°. This can be interpreted according to the one dimensional model of Broers and Schenke (Broers and and Schenke 1967) where this effect can be observed due limited thickness of the electrolyte. This model is based on the penetration depth of a concentration wave from the excitation electrode into the electrolyte. At medium frequencies the penetration depth of this perturbation wave is much smaller than the thickness of the electrolyte. The charge transport remains limited

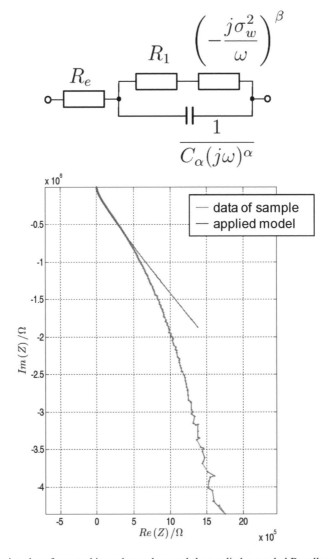

Figure 4.   Nyquist-plot of spectral impedance data and the applied extended Randles Circuit.

by the penetration depth resulting in the 45° -graph predicted by the Warburg impedance. As the perturbation wave reaches the other electrode, the end of the electrolyte film is a second limiting factor for the charge transport. If the frequency is decreased more the finite Warburg model predicts a strong increase of the imaginary part of the impedance resulting in a close to 90° slope of the graph in Figure 5. This is also represented in the equation given by Broes and Schenke:

$$Z_w = \frac{RT}{An^2F^2 c_x \sqrt{2D}} \cdot \frac{(1-j)}{\sqrt{\omega}} \cdot \coth\left(\frac{\delta}{\sqrt{2D}}(1+j)\sqrt{\omega}\right) \tag{1}$$

The constants are explained in table 1. The first two factors in this equation are also present in the regular Warburg impedance, further discussed in (Barsoukov and Macdonald 2005). The last factor is a coth function with the ratio of layer thickness δ and penetration depth as parameters. As this ratio approaches 1 the imaginary part of the Impedance starts to decrease. This corner-frequency-like effect is usable for parameter extraction even though the

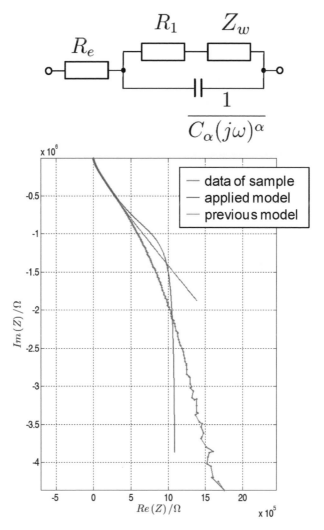

Figure 5.   Nyquist-plot of impedance data and the finite Warburg Model.

Table 1.   Obtained model parameters.

	Model parameter	Value
R_e	Serial resistance	22 Ω
R_1	Bulk resistance	40.6 kΩ
R_2	Leakage resistance	28.3 MΩ
C_α	Static capacitance	21.6 nF
α	CPE coefficient	0.98

rather abstract one dimensional model may not represent all possible charge agglomeration effects. With this second diffusion related effect which only depends on the diffusion constant and the layer thickness of the electrolyte, it is possible to distinguish and to extract the initial ion concentration c_x and the diffusion constant D. A further improvement of the model is the parallel resistor shown in Figure 6.

Table 2. Constants and dimensions.

	Parameter	Value
R	Gas constant	8.314 J/mol · K
T	Absolute Temperature	298 K
n	Charge per ion	1
F	Faraday constant	96485 C/mol
δ	Film thickness	20 nm
A	Electrode area	0.06 mm²

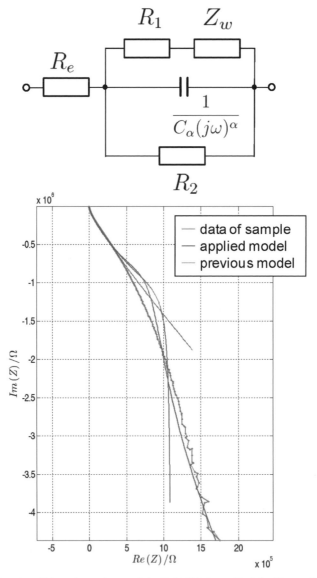

Figure 6. Nyquist-plot of impedance data and the finite Warburg Model with parallel resistor.

18

Both Warburg impedance and the capacitance in parallel result in infinite impedance values at low frequencies. A resistor parallel will limit the Impedance at low frequencies and represent all leakage currents between the electrodes. For the fitting process this increases the degree of freedom and allows better matching of the graph in the low frequency range, at the cost of reduced sensitivity for the parameters of the high frequency range. Numerical results of the fitting of the measured data with the model shown in Figure 6 are summarized in table 1. For the diffusion constant values of $D = 1.60928 \times 10^{-13}$ m^2/s and for the ion concentration values of $c_x = 1.41 \times 10^{-1}$ mol/m^3 are estimated, if the values of table 2 are used as constants during fitting of eq. 1.

5 CONCLUSION

The presented method is suitable for characterization of solid state electrolytes in thin films. The obtained parameters of electrochemical processes are the diffusion constant and the initial active ion concentration. As the major power related properties of the sample cell are related to the charge concentrations, these two parameters can represent the electrolyte's ability to agglomerate a certain number of ions in certain time. The applied multispectral analysis method can extract these parameters from the frequency range where one mechanism is dominating, even if the areas of change of the dominating mechanism might not be represented accurate by the model.

ACKNOWLEDGEMENT

The work was funded by the BMBF within the "Kompetenznetzwerk für Nanosysteminte-gration", funding No. 03IS2011, and is gratefully acknowledged.

REFERENCES

Barsoukov, E. and J.R. Macdonald (2005). *Impedance Spectroscopy Theory, Experiment, and Applications.* Wiley & Sons.

Bisquert, J. and A. Compte (2001, february). Theory of the electrochemical impedance of anomalous diffusion. *Journal of Electroanalytical Chemistry 499*(1), 112–120.

Broers, G. and M. Schenke (1967). Electrode processes in molten carbonate fuel cells. In *Abhandlung der Sächsischen Akademie der Wissenschaften zu Leipzig—Internationales Symposium Brennstoffelemente,* Volume 49, pp. 299–315. Akademie Verlag.

Büschel, P., U. Tröltzsch, and O. Kanoun (2011). Use of stochastic methods for robust parameter extraction from impedance spectra. *Electrochimica Acta 56*, 8069–8077.

Hamann, C.H. and W. Vielstich (2005). *Elektrochemie.* Wiley-VCH Verlag GmbH & Co. KGaA.

Tröltzsch, U. (2005). *Modellbasierte Zustandsdiagnose von Gerätebatterien.* Ph.D. thesis, University of the Bundeswehr.

Lecture Notes on Impedance Spectroscopy, Volume 3 – Kanoun (ed)
© *2012 Taylor & Francis Group, London, ISBN 978-0-415-64430-3*

Thermoelectrochemical investigation of silver electrodeposition from nitric and tartaric solutions

Omar Aaboubi & Abdelghani Housni
Chemistry Department, Reims Champagne Ardenne University, Reims, France

ABSTRACT: Voltammetric and thermoelectrochemical (TEC) transfer function measurements have been carried out to study the electrodeposition of silver from nitric and tartaric solutions. For an isothermal cell, the observed increase of the limiting current is due to the diffusion coefficient increase and to the mass transport boundary layer decrease when bath temperature increases. In a non-isothermal cell, through the use of sine wave temperature modulation, the TEC transfer function measurements show a typical mass transport responses and typical adsorption relaxation in middle frequency domain. The experimental data are in good accordance with previously developed model and permit to determine the diffusion activation energy and the densification coefficients of silver ions in this media.

Keywords: silver, temperature, electrochemical impedance, thermoelectrochemistry, electrode-position, adsorption

1 INTRODUCTION

A renewal of interest toward temperature effects on the interfacial electrochemical processes has been observed in the last decade, (Gründler, Kirbs, and Dunsch 2009, Burstein, Carboneras, and Daymond 2010, Aaboubi, Merienne, Amblard, Chopart, and Olivier 2002, Pasquale, Marchiano, and Arvia 2002a, Citti, Aaboubi, Chopart, Merienne, and Olivier 1996, Baranski 2002, Citti, Aaboubi, Chopart, Gabrielli, Olivier, and Tribollet 1997, Jasinski, Gründler, Flechsig, and Wang 2001, Rassaei, Compton, and Marken 2009). A significant attention has been devoted to the heated electrode and its temperature modification, either for stationary or dynamic regimes, (Citti, Aaboubi, Chopart, Merienne, and Olivier 1996, Baranski 2002, Citti, Aaboubi, Chopart, Gabrielli, Olivier, and Tribollet 1997, Jasinski, Gründler, Flechsig, and Wang 2001, Rassaei, Compton, and Marken 2009, Valdes and Miller 1988, Olivier, Merienne, Chopart, and Aaboubi 1992, Gabrielli, Keddam, and Lizee 1993, Rotenberg 1997, Smalley, Geng, Chen, Feldberg, Lewis, and Cali 2003, Aaboubi, Citti, Chopart, Gabrielli, Olivier, and Tribollet 2000), by modulating of the electrode temperature, (Valdes and Miller 1988, Olivier, Merienne, Chopart, and Aaboubi 1992, Gabrielli, Keddam, and Lizee 1993, Rotenberg 1997), or using temperature jump techniques, (Smalley, Geng, Chen, Feldberg, Lewis, and Cali 2003). In the published works, several devices have been developed for heating electrode and have shown their abilities to characterize the electrochemical process or for analytical purpose (Gründler, Kirbs, and Dunsch 2009, Jasinski, Gründler, Flechsig, and Wang 2001). Grundler et al., wrote a notable review in the field and reported that thermoelectrochemistry can be considered to be the most important technology, (Gründler, Kirbs, and Dunsch 2009).

In previous works we have shown the great importance to control the working electrode temperature (non-isothermal process) and developed a new transfer function based on the sine wave modulation of the electrode temperature, (Olivier, Merienne, Chopart, and Aaboubi 1992). Thus the thermoelectrochemical (TEC) transfer function has been experimentally measured and compared with theoretical models for mass transport controlled systems or charge transfer controlled system involving adsorbed species (i.e., nickel electrodeposition from Watts solution), (Aaboubi, Merienne, Amblard, Chopart, and Olivier 2002). In addition, the use of a superimposed thermal gradient should be able to give further information about the whole cathodic process, since the TEC method has revealed its ability, as well as a technique of mass transport analysis, as for the determination of characteristic parameters of adsorption or kinetic process, (Aaboubi, Merienne, Amblard, Chopart, and Olivier 2002).

In this work we report experimental results on silver electrodeposition from nitric and tartaric solution, where diffusion, charge transfer and adsorption processes are occurred, (Amblard, Froment, Georgoulis, and Papanastasiou 1978, Zarkadas, Stergiou, and Papanastasiou 2005, Pasquale, Marchiano, and Arvia 2002b, Aaboubi, Amblard, Chopart, and Olivier 2004). As it was previously shown during silver electrodeposition tartaric acid may be used as a dendritic inhibitor with a good performance as cyanide bath but without their ecological or health problems (Zarkadas, Stergiou, and Papanastasiou 2005).

2 EXPERIMENTAL

In isothermal conditions, a classical three electrode cell was used where, the working electrode (WE) was a silver disk ($\varnothing = 4$ mm). To avoid any salts precipitation, two silver bars of great area were used one as the reference and the other as the counter electrode. The non-isothermal cell (NITC) used in this work has been previously described in (Citti, Aaboubi, Chopart, Gabrielli, Olivier, and Tribollet 1997).

Prior to each experiment, the WE was recovered by a thin silver redeposit was electroplated at E_{ap} = 0.33 V/SCE for 2 minutes. The chemical composition of electrolytic solution was KNO_3 (0.5 mol/dm^3), $C_4H_6O_6$ (0.015 mol/dm^3) and different concentrations of $AgNO_3$ at pH = 2.45. The temperature of the solution T_b was maintained constant by circulation water in a double-jacketed cell.

The polarization curves were measured using a potentiostat/galvanostat (Radiometer) and potential scan rate of $v = 0.25$ mV/S. The TEC measurements were carried out using a home-made potentiostat and a frequency response analyser (Solartron 1250) for frequency range from 20 Hz to 1 mHz. The NITC was provided with two Zener diodes lacked on the back face of the WE. One of them was used to heat the electrode and the other to measure its temperature. After the calibration of the diode, its voltage, leads to the electrode temperature, T_e, with good accuracy and the so-called thermal gradient, $(\Delta T = T_e T_b)$ can be evaluated. The sine wave modulation of T_e could be made with a great accuracy for frequencies up to 10 Hz. In this work the temperature modulation amplitude of $T = 0.7$ K$_{rms}$ was used.

3 RESULTS AND DISCUSSION

3.1 *Isothermal cell*

Using the isothermal cell, the polarization curves $I(E)$ were measured for different bath temperatures T_b and various $AgNO_3$ concentrations, C^*. The data reported in figure 1 clearly show a strong enhancement of the electrolysis current as temperature increases. For high polarization values the diffusion process predominates and the current increase may be related to the diffusion coefficient and the thickness of diffusion layer modification with temperature. Figure 2 shows the equilibrium potential, E_{eq}, evolution as function of bath temperature. The data shows a linear evolution of $E_{eq} = f(T_b)$ for the various $AgNO_3$ concentration values. The slope of the straight line corresponds to the temperature coefficient of Ag^+/Ag system in this media. A value of $dE_{eq}/dT = (0.38 \pm 0.08)$ mV/K has been obtained.

For mass transport controlled system under isothermal conditions, the diffusion limiting current I_L may be expressed as follow

$$I_L = -nFSC^*m \qquad (1)$$

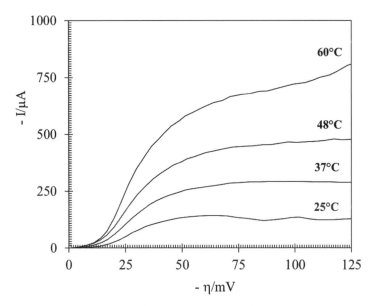

Figure 1. Experimental current-overpotential curves plotted for various bath temperature values. $[AgNO_3] = 0.025$ mol/dm³; $[KNO_3] = 0.5$ mol/dm³; $[C_4H_6O_6] = 0.015$ mol/dm³.

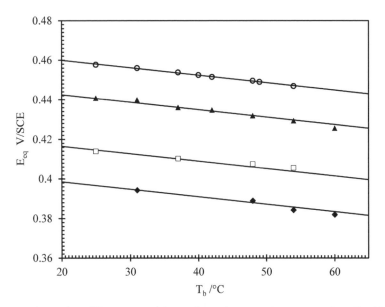

Figure 2. Experimental equilibrum potential vs T_b for various $AgNO_3$ concentration $C^* = (\circ)$ 50 mmol/dm³; (▲) 25 mmol/dm³; (□) 10 mmol/dm³; (◆) 5 mmol/dm³.

where only the mass transport coefficient, $m = m/\delta$, is depending on the temperature. Usually the diffusion coefficient can be expressed as an exponential function of the temperature (Gründler, Kirbs, and Dunsch 2009) and the $D(T)$ relationship is written as follows

$$D = D_0 \exp\frac{-E_a}{RT} \tag{2}$$

where E_a is the activation energy of diffusion process and D_0 is the pre-exponential factor. Since the mass transport coefficient can be calculated from the experimental value of the

23

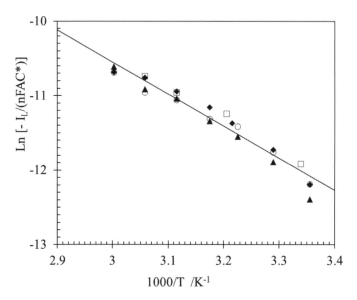

Figure 3. Evolution of ln $I_L/nFSC*$ v.s. $1/T$ for various $AgNO_3$ concentration values. $C* = $ (o) 50 mmol/dm³; (▲) 25 mmol/dm³; (□) 10 mmol/dm³; (◆) 5 mmol/dm³. All parameters are expressed in MKSA units.

limiting current (see Eq. 3) and may be also expressed in the form of an Arrhenius equation as follows

$$m = \frac{-I_L}{nFSC^*} = m_0 \exp \frac{-W}{RT} \qquad (3)$$

where W is an equivalent activation energy and m_0 is the pre-exponential factor. In Eq. 3 S is the surface area and the other parameters have the usual meaning.

In Figure 3, a typical Arrhenius plot, $\ln - I_L/nFSC^*$ against $(1/T)$ is obtained for different $AgNO_3$ concentration values. From the straight line, the values of $m_0 = 10.7$ m/s and $W = 35.8$ kJ/mol have been obtained. However, as the temperature dependency of isothermal diffusion layer $\delta(T)$ is not well known, we cannot evaluate directly, D_0 and E_a. We will see afterwards how the use of the TEC transfer function permits to circumvent this difficulty.

3.2   Non-isothermal cell

Figure 4 shows typical experimental evolution of the limiting current as function of the thermal gradient, ΔT for various $AgNO_3$ concentrations. As already shown in the case of mass transport controlled system (Pasquale, Marchiano, and Arvia 2002a, Citti, Aaboubi, Chopart, Merienne, and Olivier 1996, Pasquale, Marchiano, and Arvia 2002b), the increase of the limiting current may be related to the thermo convective effect generated by applying the thermal gradient near the electrode. It is worth noting that due to the presence of tartaric acid no drift of the current values has been observed. Hence, the use of tartaric acid as an inhibitor agent avoids the working electrode thickness development during the experiment.

According to Marcchiano & Arvia, model (Marchiano and Arvia 1968) developed in the case of laminar thermal convection occurring at a vertical heated electrode of diameter d, the diffusion limiting current I_L defined in (Citti, Aaboubi, Chopart, Merienne, and Olivier 1996, Marchiano and Arvia 1968) may be expressed in the following form:

$$f - f_0 = X^3 \left(\frac{\beta_0^{3/4}}{Pr^{1/4}} \right) \left(\frac{\Delta T}{C^*} \right)^{3/4} \qquad (4)$$

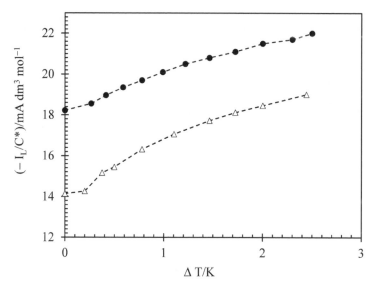

Figure 4. Experimental cathodic limiting current $-I_L/C^*$ vs T. (●) $C^* = 100$ mmol/dm³ and (Δ) $C^* = 50$ mmol/dm³.

where $\beta_0 = (1/\rho^*)(\partial\rho/\partial T)$ is the volumetric expansion coefficient, $\mathrm{Pr} = v/a$, is the Prandtl number (a the thermal diffusivity and v is the kinematics viscosity) and the parameters X, f and f_0 are defined as follow

$$X = 0.77nFD^{2/3}\left(\frac{g}{4v^{2/3}}\right)^{1/4}d^{7/4}$$

$$f = C^{*3/4}\left(\frac{I_L}{C^*}\right)^3$$

$$f_0 = X^3\left(\frac{\alpha_0^{3/4}}{\mathrm{Sc}^{1/4}}\right)$$

(5)

In Eq. 5, g is the gravity acceleration, $\mathrm{Sc} = v/D$ is the Schmidt number, $\alpha_0 = (1/\rho)(\partial\rho/\partial C)$ is the specific densification coefficient, C is the concentration and ρ is the density of the solution.

In figure 5, all the experimental data can be reduced to one curve by considering the evolution of $(f-f_0)$ as function of $(T/C^*)^{3/4}$. From the straight line the values of the specific densification coefficient, α_0 and the volumetric expansion coefficient, β_0 have been obtained. The values reported in table 1 are in good agreement with data previously reported in (Citti, Aaboubi, Chopart, Merienne, and Olivier 1996, Aaboubi, Citti, Chopart, Gabrielli, Olivier, and Tribollet 2000, Pasquale, Marchiano, and Arvia 2002b).

3.3   TEC transfer function measured at the limiting current

Figure 6 shows a typical TEC transfer function diagrams measured at the diffusion limiting current. The general shape of the diagrams depends on the value of the thermal gradient and $AgNO_3$ concentrations. In most cases the measurements exhibit a good reproducibility and dispersion less than 5%. In Figure 6, the low frequency loop is related to the mass transport process and the accumulation real point is related to the diffusion coefficient modifications due the temperature modulation. According to Rotenberg the high frequency branch may be related to the modification of the space charge in the double layer due to the temperature modulation (Rotenberg 1997). The diagrams shape is the same as that obtained

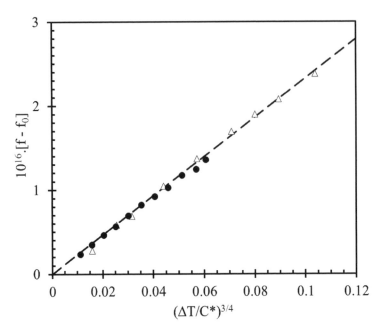

Figure 5.   Evolution of $[f - f_0]$ v.s. $(Z/C^*)^{3/4}$. Same conditions as in Figure 4. (•) $C^* = 0.1$ mol/dm³ and (Δ)$C^* = 0.05$ mol/dm³. All the parameters are expressed in MKSA units.

Table 1.   Comparison of the densification coefficient α_0 and volumetric expansion coefficient β_0 values deduced from TEC data and the stationary limiting current measurements for two $AgNO_3$ concentration values.

	0.05 mol/l		0.1 mol/l	
	$10^5 \times \alpha_0$ (m^3/mol)	$10^4 \times \beta_0$ (1/K)	$10^5 \times \alpha_0$ (m^3/mol)	$10^4 \times \beta_0$ (1/K)
TEC transfer function measurement	3.7 ± 0.5	4.1 ± 0.5	3.2 ± 0.3	3.9 ± 0.6
Stationary measurement	3.0 ± 0.3	0.1	3.6 ± 0.3	3.3 ± 0.1

by using a reversible electrochemical system limited by mass transport e.g. potassium ferri-ferrocyanide (Aaboubi, Citti, Chopart, Gabrielli, Olivier, and Tribollet 2000). According to previous studies, the obtained loops start at the origin around 0.5 Hz and finish all the faster as ΔT increases with an amplitude all the smaller as ΔT increases. Such behaviour results from the decrease of the diffusion layer thickness of the due to the enhancement of the thermal convection.

According to (Aaboubi, Citti, Chopart, Gabrielli, Olivier, and Tribollet 2000) the total TEC transfer function is equal to the sum of tree terms

$$Y = \frac{\Delta I}{\Delta T} = j\omega\left(\frac{\partial q}{\partial T}\right) + \left(\frac{I}{D}\right)\left(\frac{\partial D}{\partial T}\right) + Y_C \qquad (6)$$

In which the first term corresponds to the charge of the interface, the second term expresses the temperature dependency of the diffusion coefficient and YC is the TEC mass transport transfer function. According to the theory developed by Aaboubi et al. (Aaboubi, Citti,

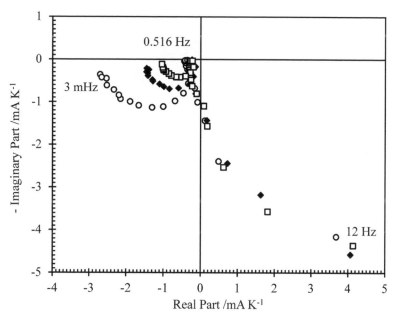

Figure 6. TEC transfer function diagrams measured at limiting current for silver electrodeposition.
(o) $\Delta T = 0.7$ K; ($\blacklozenge$) $\Delta T = 1.3$ K; ($\square$) $\Delta T = 2.4$ K. $C^* = 0.1$ mol/dm³ and $T_b = 298$ K.

Chopart, Gabrielli, Olivier, and Tribollet 2000), the transfer function Y_C is defined by the following expression

$$Y_C = \left(\frac{0.374nFDC^*}{A} \left[\frac{\beta_0^{3/4}}{\Delta T^{1/4} Pr^{1/4}} \right] (ASc)^{1/3} \left(\frac{g}{4\nu^2} \right)^{1/4} \right) H(\sigma) \qquad (7)$$

The analytic expressions of the transfer function $H(\sigma)$ is given in (Aaboubi, Citti, Chopart, Gabrielli, Olivier, and Tribollet 2000) as a function of the dimensionless frequency $\sigma = \omega \delta_{th}^2 / D$ where δth corresponds to the diffusion layer thickness imposed by the thermal gradient. In Eq. 7 the function A is defined in (Marchiano and Arvia 1968).

Figure 7 shows the experimental TEC transfer function plotted in Bode plane. The plotting against the dimensionless frequency σ shows that all the diagrams are superposed. For each experiment, the comparison of the experimental data to the calculated TEC transfer function (Eq. 7) (Aaboubi, Citti, Chopart, Gabrielli, Olivier, and Tribollet 2000) may lead to the values of the specific densification and thermal expansion coefficient α_0 and β_0. In table 1, the comparison of the values of α_0 and β_0 deduced from the stationary measurements (Fig. 5) and that obtained from the TEC transfer function measurements shows a good accordance between the two determination methods. The obtained data are in agreement with that previously obtained by Pasquale et al., during silver electrodeposition in sulphate media (Pasquale, Marchiano, and Arvia 2002a).

In Figure 5, the accumulation real point, M_{HF}, is related to the temperature variation of the diffusion coefficient. According to the Fick's law and tacking account Eq. 2, the following equation may be obtained

$$\frac{M_{HF}}{I_L} = \left(\frac{E_a}{RT^2} \right) \qquad (8)$$

The experimental M_{HF}/I_L ratios and the deduced activation energy value are listed in table 2 for various ΔT and C^* values. The mean value of the activation energy $E_a = 11.65$ kJ/mol is close to that previously obtained for other metallic ions (Gründler, Kirbs, and Dunsch 2009).

27

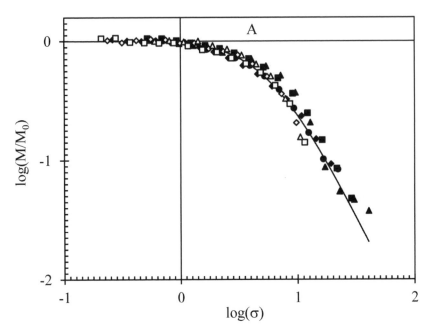

Figure 7A. Calculated and experimental $\log(M/M_0)$ vs $\log(\sigma)$ at limiting current. $C^* = 0.1$ mol/dm^{-3}: (◊) $\Delta T = 1.3$ K; (△) $\Delta T = 2$ K and (□) $\Delta T = 2.4$ K. $C^* = 0.5$ mol/dm^{-3}: (●) $\Delta T = 0.7$ K; (◆) $\Delta T = 1.3$ K; (▲) $\Delta T = 2$ K; (■) $\Delta T = 2.4$ K.

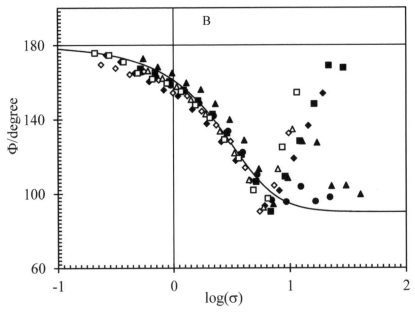

Figure 7B. Calculated and experimental phase shift, Φ vs $\log(\sigma)$ at limiting current. $C^* = 0.1$ mol/dm^{-3}: (◊) $\Delta T = 1.3$ K; (△) $\Delta T = 2$ K and (□) $\Delta T = 2.4$ K. $C^* = 0.5$ mol/dm^{-3}: (●) $\Delta T = 0.7$ K; (◆) $\Delta T = 1.3$ K; (▲) $\Delta T = 2$ K; (■) $\Delta T = 2.4$ K.

3.4 *Isothermal diffusion boundary layer $\delta = f(T_b)$ evolution*

From the diffusion activation energy value E_a determined by the TEC technique and the value of diffusion coefficient at 25°C given in (Hull 1928) we can now write the general expression of the diffusion coefficient (Eq. 2) and consequently calculated the isothermal

Table 2. Experimental limiting current, I_L, high frequency modulus, M_{HF} and (M_{HF}/I_L) ratio values for various thermal gradient and $AgNO_3$ concentration values.

$\Delta T(K)$	0.05 mol/l			0.05 mol/l		
	I_L (mA)	$10^2 \times R_{th1}$ (mA/K)	$10^2 \times R_{th1}/I_L$ (1/K)	I_L (mA)	$10^2 \times R_{th1}$ (mA/K)	$10^2 \times R_{th1}/I_L$ (1/K)
0.7	0.794	0.76	0.957	1.95	4.27	2.19
1.3	0.862	1.54	1.79	2.02	2.98	1.47
2.0	0.905	1.85	2.04	2.15	2.84	1.32
2.4	0.938	1.33	1.42	2.25	3.21	1.43

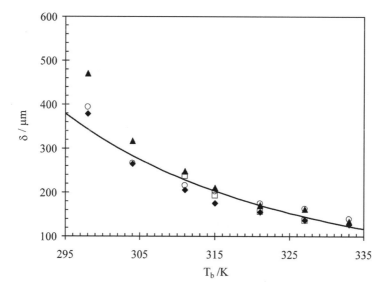

Figure 8. Calculated isothermal diffusion layer δ vs T_b. Same conditions as in Figure 3. The line corresponds to the calculated δ from Eq. 12.

diffusion boundary layer δ using the experimental data reported in figure 3. The variation of δ vs. T_b given in figure 7 exhibits an exponential type law in agreement with the tendency curve expressed by

$$\delta = 0.0202 \exp\left(\frac{2903.5}{T}\right) \qquad (9)$$

in which δ is expressed in μm and T is expressed in degree K.

3.5 TEC transfer function measured at various polarizations

Figure 9A shows the potential dependence of the TEC transfer function measured with $AgNO_3$ concentration of $C^* = 0.01$ mol/dm^3. In addition to the mass transport loop and the space charge branch a middle frequency loop is visible in the positive part of the plan. Its amplitude is as much higher as polarization increased and completely disappears at the limiting current value where the process is under pure mass transport control. The middle frequency loop points out the previous results obtained during nickel electrodeposition from Watts's bath (Aaboubi, Merienne, Amblard, Chopart, and Olivier 2002) and may be attributing to the contribution of the adsorption process during silver electrodeposition. As presented in Figure 9B, the TEC transfer function diagram present two characteristic limits, the adsorption low frequency limit, $M_1(0)$ and the mass transport low frequency limit, $M_2(0)$, each of them changing with the polarization value.

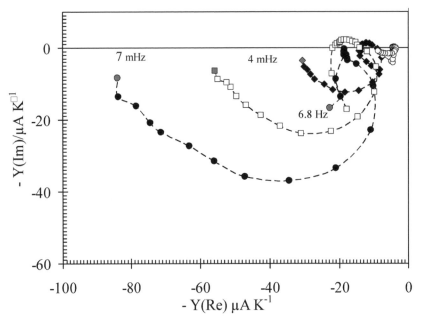

Figure 9A. TEC transfer function diagrams measured at various polarizations ΔE_p for silver electrodeposition. (O) $E_p = 0.39$ V/SCE; (♦) $E_p = 0.365$ V/SCE; (□) $E_p = 0.34$ V/SCE and (●) $E_p = 0.29$ V/SCE. $\Delta T = 2$ K; $C^* = 0.01$ mol/dm^{-3} and $T_b = 298$ K.

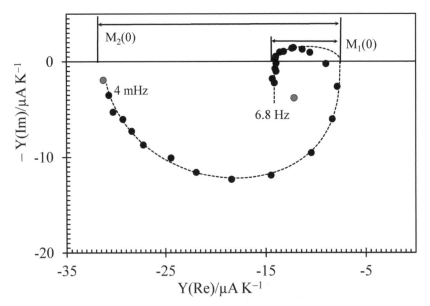

Figure 9B. TEC transfer function diagrams measured for silver electrodeposition at $E_p = 0.365$ V/SCE. $\Delta T = 2$ K; $C^* = 0.01$ mol/dm^{-3} and $T_b = 298$ K.

3.6 Adsorption low frequency limit

In agreement with the theory developed by Aaboubi et al. (Aaboubi, Merienne, Amblard, Chopart, and Olivier 2002), the low frequency limit of the adsorption response can be written as follow

$$M_1(0) = I \frac{A_1(T) + A_2(T)}{1 + \frac{R_s}{R_{ct}} + \frac{I}{2}A_3(T)} \tag{10}$$

In eq. 10, I is the electrolysis current, R_s is the solution resistance, R_{ct} is the charge transfer resistance and the function $A_n(T)$ are defined in (Aaboubi, Merienne, Amblard, Chopart, and Olivier 2002). As already shown in Figure 10A, by plotting $M_1(0)$ against I, a linear relationship was found for the two $AgNO_3$ concentration.

3.7 Mass transport low frequency limit

Assuming that the adsorption process may be neglected and the Ag(I) ions discharge is under mass transport and charge transfer control, the stationary faradic current may be expressed by the Butler-Volmer equation

$$I = -nFSCk_0 \exp\frac{b}{T}\eta \tag{11}$$

where $b = -\alpha n F/R$, α is the charge transfer coefficient, $\eta = E - E_{eq} - R_s I$ is the overpotential and k_0 is the rate constant at the equilibrium potential. In addition, when the frequency tends toward zero, the low frequency limit of the TEC transfer function corresponds to the derivative of stationary current as function of temperature. Therefore, from eq. 11 the expression of $M_2(0)$ is

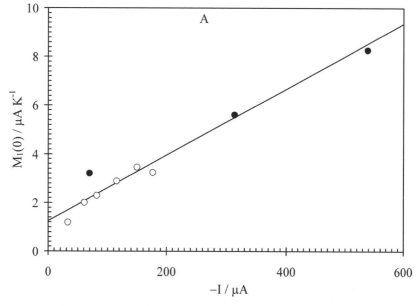

Figure 10A. Evolution of adsorption low frequency limit $M_1(0)$ vs $-I$ (stationary cathodic current). (○) $C^* = 0.01$ mol/dm^{-3} and (●) $C^* = 0.05$ mol/dm^{-3}. $\Delta T = 2$ K; and $T_b = 298$ K.

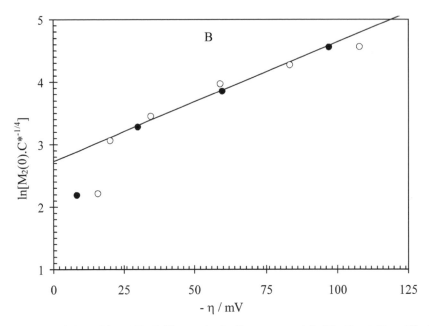

Figure 10B. Evolution of ln $M_2(0)$ $C^{*-1/4}$ vs $-\eta$ (cathodic overpotential). (O) $C^* = 0.01$ mol/dm^{-3} and (●) $C^* = 0.05$ mol/dm^{-3}. $\Delta T = 2$ K; and $T_b = 298$ K.

$$M_2(0) = M_0 \exp\frac{b}{T}\eta \qquad (12)$$

where

$$M_0 = -nFSCk_0\left[\frac{dC}{dT} + \frac{b}{T}\frac{d\eta}{dT} - \frac{b\eta}{T^2}\right] \qquad (13)$$

In figure 10B the values of ln($M_2(0)$) against the cathodic overpotential $-\eta$ are plotted. It is obvious that for the two $AgNO_3$ concentrations, a linear variation is obtained over more than 100 mV range in agreement with eq. 12.

4 CONCLUSION

From these results it comes out that: During silver electrodeposition from nitric solution, the useof small quantity of tartaric acid permits to carry out the TEC transfer function the without fearing the appearance of dendrite growths. In isothermal cell, the increase of the diffusion limiting currents with bath temperature is due to the increase of diffusion coefficient and to the decrease diffusion boundary layer.

In non-isothermal cell, the increase of the limiting currents is related to a thermal convection generated by applying the thermal gradient in the vicinity of the electrode. At the limiting current polarization the diagrams is composed of charges space branch and a mass transport loop. The evolution of this loop is in accordance with the previously developed model for a pure mass transport controlled system. The evolution of the high frequency limit against the thermal gradient values permits to get the diffusion activation energy and consequently to quantify the temperature evolution law of the thickness of the isothermal diffusion layer. In addition, for polarization value far from the limiting current a typical adsorption loop appears in the middle frequency which has a Tafel's type evolution as function of the over potential. Moreover, the investigation of the diffusion and adsorption responses allowed us to show that the use of the TEC transfer function measurements may be generalized to the study of various electrochemical processes.

REFERENCES

Aaboubi, O., J. Amblard, J. Chopart, and A. Olivier (2004). Magnetohydrodynamic analysis of silver electrocrystallization from a nitric and tartaric solution. *Journal of the Electrochemical Society 151*, C112.

Aaboubi, O., I. Citti, J. Chopart, C. Gabrielli, A. Olivier, and B. Tribollet (2000). Thermoelectrochemical transfer function under thermal laminar free convection at a vertical electrode. *Journal of the Electrochemical Society 147*, 3808.

Aaboubi, O., E. Merienne, J. Amblard, J. Chopart, and A. Olivier (2002). A thermoelectrochemical transfer function analysis of nickel electrocrystallization in a nonisothermal cell. *Journal of the Electrochemical Society 149*, E90.

Amblard, J., M. Froment, C. Georgoulis, and G. Papanastasiou (1978). Sur le rôle de l'acide tartrique et des tartrates dans l'inhibition de la croissance dendritique de l'argent électrodéposé à partir d'une solution aqueuse de nitrate. *Surface Technology 6*(6), 409–423.

Baranski, A. (2002). Hot microelectrodes. *Analytical chemistry 74*(6), 1294–1301.

Burstein, G., M. Carboneras, and B. Daymond (2010). The temperature dependence of passivity breakdown on a titanium alloy determined by cyclic noise thermammetry. *Electrochimica Acta 55*(27), 7860–7866.

Citti, I., O. Aaboubi, J. Chopart, C. Gabrielli, A. Olivier, and B. Tribollet (1997). Impedance of laminar free convection and thermal convection at a vertical electrode. *Journal of the Electrochemical Society 144*, 2263.

Citti, I., O. Aaboubi, J. Chopart, E. Merienne, and A. Olivier (1996). Thermoelectrochemical impedance (tec)–ii. validation of an electrochemical mass transport-controlled system by stationary and dynamic investigations. *Electrochimica Acta 41*(17), 2731–2736.

Gabrielli, C., M. Keddam, and J. Lizee (1993). Frequency analysis of a temperature perturbation technique in electrochemistry: Part i. theoretical aspects. *Journal of Electroanalytical Chemistry 359*(1–2), 1–20.

Gründler, P., A. Kirbs, and L. Dunsch (2009). Modern thermoelectrochemistry. *ChemPhysChem 10*(11), 1722–1746.

Hull, C. (1928). *International critical tables of numerical data, physics, chemistry and technology*, Volume 3. Pub. for the National research council by the McGraw-Hill book company, inc.

Jasinski, M., P. Gründler, G. Flechsig, and J. Wang (2001). Anodic stripping voltammetry with a heated mercury film on a screen-printed carbon electrode. *Electroanalysis 13*(1), 34–36.

Marchiano, S. and A. Arvia (1968). Diffusional flow under non-isothermal laminar free convection at a thermal convective electrode. *Electrochimica Acta 13*(7), 1657–1669.

Olivier, A., E. Merienne, J. Chopart, and O. Aaboubi (1992). Thermoelectrochemical impedances–i. a new experimental device to measure thermoelectrical transfer functions. *Electrochimica Acta 37*(11), 1945–1950.

Pasquale, M., S. Marchiano, and A. Arvia (2002a). Non-isothermal ionic mass transfer at vertical parallel plate electrodes under natural convection.: Comparison and validity range of dimensionless correlations. *Electrochimica Acta 48*(2), 153–163.

Pasquale, M., S. Marchiano, and A. Arvia (2002b). Transitions in the growth mode of branched silver electrodeposits under isothermal and non-isothermal ionic mass transfer kinetics. *Journal of Electroanalytical Chemistry 532*(1–2), 255–268.

Rassaei, L., R. Compton, and F. Marken (2009). Microwave-enhanced electrochemistry in locally superheated aqueous- glycerol electrolyte media. *The Journal of Physical Chemistry C 113*(8), 3046–3049.

Rotenberg, Z. (1997). Thermoelectrochemical impedance. *Electrochimica Acta 42*(5), 793–799.

Smalley, J., L. Geng, A. Chen, S. Feldberg, N. Lewis, and G. Cali (2003). An indirect laser-induced temperature jump study of the influence of redox couple adsorption on heterogeneous electron transfer kinetics. *Journal of Electroanalytical Chemistry 549*, 13–24.

Valdes, J. and B. Miller (1988). Thermal modulation of rotating disk electrodes: steady-state response. *The Journal of Physical Chemistry 92*(2), 525–532.

Zarkadas, G., A. Stergiou, and G. Papanastasiou (2005). Influence of citric acid on the silver electrodeposition from aqueous agno3 solutions. *Electrochimica Acta 50*(25–26), 5022–5031.

Lecture Notes on Impedance Spectroscopy, Volume 3 – Kanoun (ed)
© 2012 Taylor & Francis Group, London, ISBN 978-0-415-64430-3

Acquisition of impedance and gravimetric data for the characterization of electrode/electrolyte interfaces

Minghua Huang, John B. Henry & Balázs B. Berkes
Center for Electrochemical Sciences, Ruhr-Universität Bochum, Germany

Artjom Maljusch
Analytische Chemie, Elektroanalytik & Sensorik, Ruhr-Universität Bochum, Germany

Wolfgang Schuhmann
Center for Electrochemical Sciences, Analytische Chemie, Elektroanalytik & Sensorik, Ruhr-Universität Bochum, Germany

Alexander S. Bondarenko
Center for Electrochemical Sciences, Ruhr-Universität Bochum, Germany

ABSTRACT: The acquisition and analysis of electrochemical impedance spectroscopy (EIS) and electrochemical quartz crystal microbalance (EQCM) data in one cyclic potential scan allows a detailed characterization of complex non-stationary electrode/electrolyte interfaces. Analysis of the EIS-EQCM data-set enables efficient elucidation of physical models of studied systems and determination of their physico-chemical parameters. The underpotential deposition of Pb atomic layers on gold electrodes modified with an atomic layer of Ag has been used as a model system to demonstrate the EIS-EQCM approach.

Keywords: electrochemical impedance spectroscopy, electrochemical quartz crystal microbalance, underpotential deposition

1 INTRODUCTION

Investigation of the electrode/electrolyte boundary involves the deployment of a wide arsenal of electrochemical and non-electrochemical techniques (Zoski 2007). So-called direct current (DC) techniques, such as voltammetric techniques remain the primary techniques used in the characterization of non-stationary systems. Non-stationarity, in this context, means that the studied systems do not display the same properties in cycle-to-cycle sequences, or even from the forward to backward potential scans (Bondarenko, Ragoisha, Osipovich, and Streltsov 2006, Ragoisha and Bondarenko 2004). Such non-stationarity can arise from factors, such as surface alloying and/or surface reconstruction. Under these circumstances the ability of voltammetric techniques to characterize systems can be limited. Detailed characterization of these systems should, ideally, be performed within a single potential scan. To achieve this it is important to utilize complementary in-situ techniques in parallel with voltammetric techniques. If possible, these complementary techniques should allow acquisitionof the maximum amount of independent and self-consistent data from the minimum number of measurements.

In this work, it will be demonstrated that simultaneous acquisition and analysis of impedance and gravimetric data in a single cyclic potential scan (Fig. 1) enables a detailed electrochemical characterization of dynamic non-stationary electrode/electrolyte interfaces. The combination of electrochemical impedance spectroscopy (EIS) and electrochemical quartz crystal microbalance (EQCM) techniques is used to characterize the underpotential

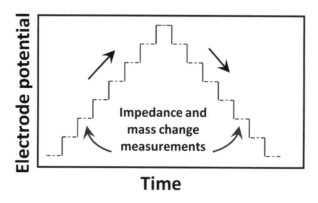

Figure 1. Combined EIS-EQCM measurements used to study non-stationary interfaces. At each of the steps of the staircase scan, EIS spectra are taken while EQCM measurements run continuously in parallel.

deposition (UPD) of sub-monolayer amounts of Pb onto Auelectrode modified with an atomic overlayer of Ag (Ag_{ad}/Au). The electrode potential, E, controls the surface status of electrode/electrolyte interfaces. Scanning of the electrode potential is therefore vital to continuously monitor theinterface at each stage of the dynamic process. The combination of EIS and EQCM measurements also makes it easier to elucidate self-consistent physical models (equivalent circuits) of the interface with varying electrode potentials providing a deeper physical insight of the non-stationary interface.

2 RATIONALE FOR A COMBINED EIS-EQCM TECHNIQUE

As DC techniques alone are not always appropriate to fully characterize non-stationary systems, it is necessary to consider the applicability of other techniques. The EIS technique involves measuring the response of the electrode interface to an alternating current (AC) probing signal with varying frequency at a range of electrode potentials (Bondarenko, Ragoisha, Osipovich, and Streltsov 2006, Ragoisha and Bondarenko 2004). Analysis of the EIS data then enables decomposition of the AC response into components related to the simultaneous processes occurring at the electrode/electrolyte interface (e.g., double layer capacitance, charge transfer resistance and diffusional properties of the system). While EIS has been shown to allow characterization of non-stationary systems (Bondarenko, Ragoisha, Osipovich, and Streltsov 2006, Pettit, Goonetilleke, Sulyma, and Roy 2006), some complex electrochemical systems can introduce ambiguities that hinder the analysis process. It becomes vital to employ other complementary techniques in conjunction with EIS to extract further information about the status of the interface and, hence, allow characterization.

With the ability to sense changes in electrode mass within the nanogram range, EQCM allows electrochemical studies to become more comprehensive, as it is now possible to detect mass changes on the electrode surface which do not contribute to the voltammetric response (Sabot and Krause 2002). Additionally, DC techniques, such as voltammetric techniques, can still quantify the total charge which passes through the interface. The combination of key elements of impedance spectroscopy, electrogravimetry and direct currenttechniques therefore unlocks the ability to characterize complex electrochemical interfaces.

To maximize the effectiveness of such a complementary approach the data acquisition should take place in a single cyclic (staircase, see Figure 1) potential scan. The multidimensional data acquired in such a combined EIS-EQCM scan, once analyzed, can provide a more accurate and detailed model for characterization of the interface. This provides a deeper physical insight into the electrode/electrolyte interface. This rationale for combining EIS, EQCM and DC techniques is summarized in Figure 2.

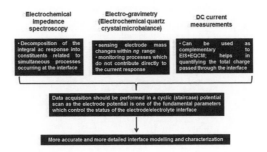

Figure 2. Schematic showing how the key elements from EIS, EQCM and DC techniques can be combined to allow a detailed characterization of non-stationary interfaces.

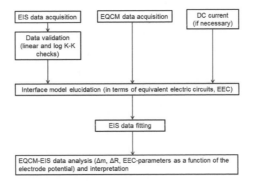

Figure 3. Protocol for data acquisition and analysis in the complementary EIS-EQCM technique.

Figure 3 shows the protocol for the data acquisition and analysis of the combined EIS-EQCM technique. Following acquisition and validation of the EIS data (Kramers-Kronig tests), EQCM, and DC data, a suitable physical model which is completely consistent with all of the complementary data is elucidated. Subsequent fitting of the EIS data to this physical model and further analysis of the available EIS-EQCM datasets enables the identification of a number of important parameters characterizing the electrode dynamics, such as: (a) the electrode mass variation, Δm (Δf); (b) the resonance resistance, ΔR, corresponding to the dissipation of the oscillation energy due to growing structures and the nature of the medium at the interface; (c) the double layer capacitance of the interface; (d) the adsorption capacitances and (e) the apparent rate coefficients of adsorption $k_f f(\Theta)$ (will be discussed later).

3 APPLICATION

Figure 4 shows the resulting data-set from the combined EIS-EQCM study of Pb under-potential deposition (UPD) on Ag_{ad}/Au in 0.1 M $HClO_4$. Preparation and characterization of the Ag_{ad}/Au surface were reported previously by Berkes et al. (Berkes, Maljusch, Schuhmann, and Bondarenko 2011). Figures 4a and 4b show the extended Bode plots for the EIS measurement of the Pb UPD process. The onset of Pb UPD can be observed at around –0.55 V (MMS) in the cathodic scan. The classic adsorption model was elucidated to be the most appropriate model for the Pb UPD process (Fig. 4c). Figure 4d illustrates the quality of fit of the experimental data to the selected equivalent circuit at selected electrode potentials. Fitting of the EIS data was performed with software reported in (Pomerantsev 2005).

The variation of the electrode mass, $\Delta m(E)$, during the potential scan is shown in Figure 4e. The EQCM measurements run in parallel with the EIS measurements. Once again the onset of the UPD is shown at around –0.55 V by the increase in electrode mass as the electrode is

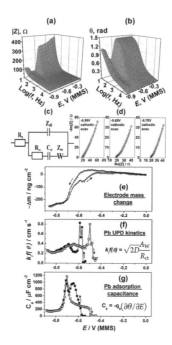

Figure 4. Analysis of the EIS-EQCM data for Pb UPD on Ag_{ad}/Au from 0.1 M $HClO_4$ and 1 mM Pb^{2+} solution, $dE/dt = 0.25$ mV/s. Extended Bode plots showing changes of (a) $|Z|$ and (b) phase shift Θ with AC frequency (30 kHz–10 Hz) and electrode potential. (c) Equivalent circuit used to fit EIS data. (d) Fitting of the EIS data to the equivalent circuit. (e) Variation of electrode mass with electrode potential. (f) Calculated kinetics of Pb UPD process on the Ag_{ad}/Au electrode as a function of electrode potential. (g) Adsorption capacitance of Pb UPD process on the Ag_{ad}/Au electrode as a function of electrode potential. All electrode potentials are versus mercury-mercury sulfate (MMS) reference electrode.

scanned cathodically, subsequent to a slow desorption process prior to Pb UPD (Fig. 4e). It has previously been shown that the formation of a Ag atomic layer on Au is accompanied by the specific adsorption of anions, most likely ClO_4^-, (Berkes, Maljusch, Schuhmann, and Bondarenko 2011). It is possible, therefore, to hypothesize that the slow desorption prior to Pb UPD is caused by perchlorate desorption from the Ag_{ad}/Au surface. The EIS data analysis of thispotential region supports this hypothesis: an additional R-C branch parallel to the double layer capacitance in the equivalent circuit is necessary to give a suitable fit to the EIS data. The apparent molar weight for the UPD-adsorbate was found to be around 200 g/mol, which is close to the expected value for Pb of 207 g/mol. Specific co-adsorption of ClO_4^- with Pb is therefore unlikely.

Using an approach previously developed by Kerner and Pajkossy (Kerner and Pajkossy 2002), it is possible to extract information about the kinetic parameters of the UPD in relation to the electrode potential. The apparent rate coefficient of the adsorption, $k_i f(\Theta)$, is obtained from the ratio of the Warburg coefficient, A_W, and the charge transfer resistance, R_{ct}, (see inset to Figure 4f). In this case, $f(\Theta)$ is a coverage related factor which is a monotonously related changing function of Θ (for example $f(\Theta) = 1 - \Theta$ in the case of the Langmuir isotherm). This apparent rate coefficient gives the rate for which atoms deposit onto the surface of a partially covered electrode, where k_i is the rate of UPD for the adatoms deposition onto a free surface. This approach has the advantage that the electrode surface area and surface concentrations of electroactive species are negated from the calculations.

The variation of $k_i f(\Theta)$ in the cyclic electrode potential scan for the Pb UPD process on Ag_{ad}/Au in 0.1 M $HClO_4$ is shown in Figure 4f. This dependence discloses complex adsorbate-adsorbate and adsorbate-substrate interactions which accordingly change at different electrode potentials and are different in cathodic and anodic scans. The initial stages of the

UPD and the very final stages of the Pb atomic layer stripping are fast (note the peaks at around –0.6 V and –0.5 V, correspondingly in Figure 4f) while at more negative potentials the rate coefficient is much smaller, where presumably the $f(\Theta)$ term plays the dominant role.

Simultaneous analysis of $\Delta m(E)$, $k_f f(\Theta)$ and adsorption capacitance $C_a(E)$ variations (Fig. 4e–g) shows that the Pb UPD is reversible in the range between around –0.7 V and –0.88 V. There is, however, a significant irreversible component at more positive potentials (Fig. 4e–g). While $C_a(E)$ and $\Delta m(E)$ give the information about the quantity of the adsorbate reduced or oxidized at each step, $k_f f(\Theta)$ dependences characterize changes in the adsorbate-adsorbate and adsorbate-substrate interactions at different potentials.

In summary, the analysis of the combined EIS-EQCM data-set appears to be efficient in elucidating physical models of non-stationary systems and determining their physico-chemical parameters.

ACKNOWLEDGMENTS

Financial support by the EU and the state NRW in the framework of the HighTech.NRW program is gratefully acknowledged.

REFERENCES

Berkes, B., A. Maljusch, W. Schuhmann, and A. Bondarenko (2011). Simultaneous acquisition of impedance and gravimetric data in a cyclic potential scan for the characterization of nonstationary electrode/electrolyte interfaces. *The Journal of Physical Chemistry C 115*, 9122–9130.

Bondarenko, A., G. Ragoisha, N. Osipovich, and E. Streltsov (2006). Multiparametric electrochemical characterisation of te-cu-pb atomic three-layer structure deposition on polycrystalline gold. *Electrochemistry Communications 8*(6), 921–926.

Kerner, Z and T. Pajkossy (2002). Measurement of adsorption rates of anions on au (111) electrodes by impedance spectroscopy. *Electrochimica acta 47*(13–14), 2055–2063.

Pettit, C., P. Goonetilleke, C. Sulyma, and D. Roy (2006). Combining impedance spectroscopy with cyclic voltammetry: measurement and analysis of kinetic parameters for faradaic and nonfaradaic reactions on thin-film gold. *Analytical Chemistry 78*(11), 3723–3729.

Pomerantsev, A. (2005). *Progress in Chemometrics Research*. Nova Science Pub Inc.

Ragoisha, G and A. Bondarenko (2004). Potentiodynamic electrochemical impedance spectroscopy of silver on platinum in underpotential and overpotential deposition. *Surface Science 566*, 315–320.

Sabot, A and S. Krause (2002). Simultaneous quartz crystal microbalance impedance and electrochemical impedance measurements investigation into the degradation of thin polymer films. *Analytical Chemistry 74*(14), 3304–3311.

Zoski, C. (2007). *Handbook of Electrochemistry*. Elsevier Science.

Lecture Notes on Impedance Spectroscopy, Volume 3 – Kanoun (ed)
© *2012 Taylor & Francis Group, London, ISBN 978-0-415-64430-3*

Comparison of eddy current theory and Finite Element Method for metal evaluation

Rauno Gordon & Olev Märtens
Thomas Johann Seebeck Department of Electronics, Tallinn University of Technology, Tallinn, Estonia

Andrei Kolyshkin
Department of Engineering Mathematics, Riga Technical University, Riga, Latvia

ABSTRACT: We are working on non-destructive evaluation of metal quality by the electric conductivity. We use contactless method with eddy currents to probe the metal piece. Here we are testing and verifying the theoretical methods for later use in practical algorithms in fast metal conductivity evaluation. In this particular work we compare the basic eddy current theory of coil impedance with the same scenario modeled with Finite Element Method (FEM). We found very good correlation between the theory and Finite Element Method below 2 kHz.

Keywords: impedance spectroscopy, eddy-currents, skin-effect, non destructive evaluation, finite element method

1   INTRODUCTION

The application we are working for is non-destructive evaluation of metal quality by the electric conductivity. We aim to use contactless method with eddy currents to probe the metal piece. The most significant application is coin validation. Coils can be used to generate eddy currents in the metal piece and the impedance of the coil depends the material under test. The coil impedance also depends on other aspects of the measurement, like coil characteristics, positioning of the metal piece relative to the coil, position of other conductive materials in the coil proximity as well as the size and shape of the metal piece.

When the conductivity of a particular metal piece has to be evaluated, the metal is placed near a coil, the impedance of the coil is measured and from that impedance value (consisting of real and imaginary part) the conductivity of the metal piece is derived with functions previously developed through theoretical and practical R&D.

We are working on testing and verifying the theoretical methods for later use in practical algorithms in fast metal conductivity evaluation. In this particular work we compare the basic eddy current theory of coil impedance over a metal plate with the same scenario modeled with Finite Element Method (FEM) in a step-by-step approach, starting with simplest coil theory and going towards more complex practical scenarios.

2   METHODS

The theoretical equations for coil impedance calculation in air are provided by (Auld, Muennemann, and Winslow 1981) that follow from original Dodd and Deeds equations (Dodd and Deeds 1968, Dodd and Deed 1975). Consider an air-core coil located above a conducting half-space with a volumetric flaw of volume V. It is shown in (Auld, Muennemann, and Winslow 1981) that the change in impedance of the coil can be computed by the formula

$$Z_{ind} = -\frac{1}{I^2}\iiint_V \Delta\sigma E_0 \cdot E_f \, dV \tag{1}$$

Here E_0 is the electric field in the absence of the flaw, E_f is the electric field in the flawed part and $\Delta\sigma = \sigma_F - \sigma_0$, where σ_0 is the conductivity of the flawless part and σ_F is the conductivity of the flawed part. Formula 1 is successfully used in (Satveli, Moulder, Wang, and Rose 1996) for the calculation of the change in impedance of a plate with a bottom cylindrical hole (the approximate formula 1 is referred to in (Satveli, Moulder, Wang, and Rose 1996) as the layer approximation since the flaw in this case is replaced by an infinite layer with the same thickness as the thickness of the flaw). Thus, E_f is the electric field in the infinite layer used to replace the flaw.

In this paper we use the layer approximation 1 in order to compute the change in impedance of an air-core coil symmetrically located above a conducting cylinder of finite thickness d (a coin, for example). The axis of the coil coincides with the axis of the cylinder. In this case formula 1 can be written in the form:

$$\Delta Z = \frac{\omega^2 \mu_0^2 n^2 \pi \Delta\sigma}{(r_o - r_i)^2 (h_1 - h_2)^2} \int_0^\infty \frac{I(\lambda r_i, \lambda r_o)(e^{-\lambda h_2} - e^{-\lambda h_1})}{\lambda^3} \, d\lambda$$

$$\times \int_0^\infty I(ur_i, ur_o)(e^{-uh_2} - e^{-uh_1}) \frac{1}{u^2\left[(u+q)^2 - (u-q)^2 e^{-2qd}\right]}$$

$$\times \left\{\frac{u+q}{\lambda+q}\left[1 - e^{-(\lambda+q)d}\right] - \frac{u-q}{\lambda-q}e^{-2qd}\left[1 - e^{-(\lambda-q)d}\right]\right\} f(R,u,\lambda) \, du \tag{2}$$

where

$$f(R,u,\lambda) = \begin{cases} \dfrac{R}{\lambda^2 - u^2}\left[uJ_1(\lambda R)J_0(uR) - uJ_1(uR)J_0(\lambda R)\right], & \lambda \neq u \\[2ex] \dfrac{R}{2u}\left\{ur\left[J_0^2(uR) + J_1^2(uR)\right] - 2J_0(uR)J_1(uR)\right\}, & \lambda = u \end{cases} \tag{3}$$

r_o and r_i are the outer and inner radii of the coil, h_1 is the lift-off, $h_2 - h_1$ is the height of the coil, ω is the frequency and n is the number of turns in the coil.

Only imaginary part of complex impedance is compared. There is no point to compare the real parts since the real part of the impedance of the coil is zero in accordance with the theory (the coil is treated as an ideal one). The coil is considered a small one, with 1 mm*1 mm cross-section that contains 36 turns and with an inner radius of 3 mm. The imaginary parts of the impedance of the coil in air are computed using the formula 2 for an infinite plate of the thickness d and are compared with the numerical results generated by the Finite Element Method.

The Finite Element Method model is developed with COMSOL Multiphysics software using axis-symmetric with variable mesh density. The 2D model contains cross-section of the coil with the highest mesh density, cross-section of the metal sample with slightly lower mesh density and a 10 cm air box around the area with decreasing mesh density. The model contains around 22000 triangular elements and with the fine mesh it can resolve the skin effect adequately even when high frequencies of 1 MHz are used in the coil area and in the metal sample. The mesh density variation in the model is around 2000 times, with smallest elements in the edges of the coil and metal sample surface and the largest elements in the outer regions of the surrounding air box (Fig. 1).

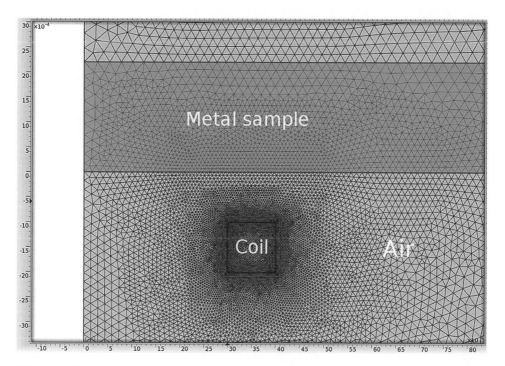

Figure 1. Finite Element Method mesh for 1 mm² coil impedance simulation. Symmetry axis is on the left of the model.

2.1 Coil in air

The theory and Finite Element Method are first used to calculate the coil impedance in air with setting the material properties for metal-area the same as air. The coil has an inner radius $r_i = 3$ mm, outer radius $r_o = 4$ mm, and height $h_1 = 1$ mm. The coil wire was considered to be with 2.5e-8 m² cross-section. The results of comparison are shown in table 1.

As can be seen from table 1, numerical results generated by Finite Element Method are in good agreement with theoretical data for a wide range of frequencies.

2.2 Coil above infinite plate

Second, the theory of Dodd and Deeds is used to compute the change in impedance of the coil located above an infinite conducting plate with the same thickness and conductivity as the thickness and conductivity of the metal piece. The lift-off of the coil from the metal sample was $h_2 - h_1 = 1$ mm. The summary is shown in table 2.

These results also show good agreement between COMSOL calculations and theory based on an infinite plate for the case where the parameters of the metal piece are $\sigma = 9.28$ MS/m, $d = 2.2$ mm, $r_c = 12.875$ mm (the radius of the metal piece). The values of the imaginary part of the change in impedance practically coincide with the results generated by Finite Element Method so that the coin can be treated as an infinite plate in this case. The induced currents in the metal sample can be calculated with the FEM simulation and the skin depth illustrated (Fig. 2).

2.3 Coil above a metal piece

For third, the theory is used in the layer approximation mode, with the flaw representing the metal piece and having the electrical conductivity of $\sigma = 9.28$ MS/m. The dimentions of the metal piece are chosen to represent a 2 coin with thickness $d = 2.2$ mm and radius

43

Table 1. The imaginary part of the impedance (in Ω).

Frequency	Theory, coil in air	Finite Element Method, coil in air
1 kHz	0.077	0.077
10 kHz	0.7705	0.7698
100 kHz	7.705	7.698
1 MHz	77.053	76.984

Table 2. The imaginary part of the impedance (in Ω).

Frequency	Theory, coil over infinite plate	Finite Element Method, coil over infinite plate
1 kHz	–0.0764	–0.0765
10 kHz	–0.6965	–0.6965
100 kHz	–6.340	–6.341

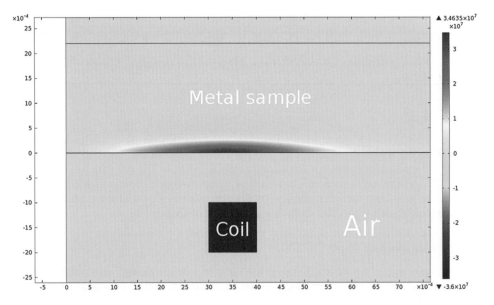

Figure 2. Induced current density (A/m²) in the metal sample at 100 kHz calculated with Finite Element Method. Symmetry axis is on the left of the model.

Table 3. The imaginary part of the impedance (in Ω).

Frequency	Theory, layer approximation	Finite Element Method, coil with metal piece
1 kHz	–0.001	–0.001
1.584 kHz	–0.002	–0.002
2.512 kHz	–0.005	–0.006
3.981 kHz	–0.011	–0.015
6.309 kHz	–0.023	–0.035
10 kHz	–0.042	–0.073

$r_c = 12.875$ mm. The comparison of the FEM and theoretical results are summarized in table 3.

This table illustrates the fact that the layer approximation seems to be working well for small frequencies. For higher frequencies the differences are becoming very noticeable already above 2 kHz.

3 CONCLUSIONS

These results allow us to use the theory for practical applications in a very wide frequency range in the "infinite plate" scenario, where the measured metal sample is much larger than the coil. For smaller coil-size metal samples we can safely use the theoretical results only on lower frequencies below 2 kHz. For higher frequencies the theoretical methods as well as the model in Finite Element Method have to be investigated and possible reasons for divergence found before proper confidence in these methods can be attained.

ACKNOWLEDGMENTS

Current work have been supported by EU (FP7-SME project "SafeMetal"), Enterprise Estonia (support of Competence Centre ELIKO), target financing SF0142737s06 and grant ETF8905 (Estonian Science Foundation) and by the European Union through the European Regional Development Fund.

REFERENCES

Auld, B., F. Muennemann, and D. Winslow (1981). Eddy current probe response to open and closed surface flaws. *Journal of Nondestructive Evaluation* 2(1), 1–21.

Dodd, C. and W. Deed (1975). Calculation of magnetic fields from time-varying currents in the presence of conductors. *NASA STI/Recon Technical Report N 76*, 19346.

Dodd, C. and W. Deeds (1968). Analytical solutions to eddy-current probe-coil problems. *Journal of Applied Physics* 39(6), 2829–2838.

Satveli, R., J. Moulder, B. Wang, and J. Rose (1996). Impedance of a coil near an imperfectly layered metal structure: The layer approximation. *Journal of Applied Physics* 79(6), 2811–2821.

Lecture Notes on Impedance Spectroscopy, Volume 3 – Kanoun (ed)
© *2012 Taylor & Francis Group, London, ISBN 978-0-415-64430-3*

Eddy current validation of Euro-coins

Rauno Gordon, Olev Märtens, Raul Land, Mart Min, Marek Rist &
Alina Gavrijaševa
Thomas Johann Seebeck Department of Electronics, Tallinn University of Technology, Tallinn, Estonia

Andrei Pokatilov
AS Metrosert, Tallinn, Estonia

Andrei Kolyshkin
Department of Engineering Mathematics, Riga Technical University, Riga, Latvia

ABSTRACT: The measurement of electrical conductivity using eddy current (electro magnetic induction) sensors is the most promising solution for validation of metal coins in coin handling devices, as the electrical conductivity is a very distinctive property of specific alloys. Low cost, simple and robust nature, and also possibility for high-speed precise measurements are the advantages of electromagnetic sensors. Multi-frequency scanning the coins under test at various field penetration depths is preferred due to the sophisticated "construction" of various alloys in nowadays coins. Air-core measurement coils have an additional advantage—a possibility to measure absolute values of conductivity without repetitive calibration of sensors and the measurement systems. Development of solutions for eddy-current validation of metal coins has been investigated together with comparing the experimental results with the values derived from theoretical models. Frequency range of 10 kHz to 10 MHz have been considered (particularly up to 500 kHz for precision measurements) and conductivity values to be measured from 4 to 40% of IACS (International Annealed Copper Standard) conductivity (about 60 MS/m). The results demonstrate that the precise, high-speed, low-cost and robust coin validation system can be developed on the basis of air-core coil arrays.

Keywords: coin validation, eddy current, inductive sensors, conductivity, coin validation, electromagnetic sensors

1 INTRODUCTION

Electrical conductivity is one of the most distinctive properties of various metal alloys (Rossiter 1991). So the electrical conductivity of metal coins consisting nowadays of sophisticated alloys and structures, are widely used for coin discrimination and validations, as proposed in (Howells 2009, Harris, Churchman, Sharman, et al., 2007). Frequently this method is used in combination with other (acoustical, imaging) sensors and corresponding signal processing (Carlosena, Lopez-Martin, Arizti, Martínez-de Guerenu, Pina-Insausti, and García-Sayés 2007). A clear model for analysis of the impedance of the measurement coil placed above the metal plate has been developed by Dodd and Deeds. Their analytical (Dodd and Deeds 1968) and numerical results (Dodd and Deed 1975) are published more than 35 years ago already. In the last work (Dodd and Deed 1975), only active (real) part of the coil impedance (losses in the measurement coil) has been considered in the calculations. In the work (Bowler and Huang 2005) by Bowler and Huang, both eddy current and contact-based (ohmic) measurement theory for metal plates has been investigated, theoretically and experimentally in parallel. More advanced numerical model for eddy-current simulations has been developed and published by Theodoulidis and his colleagues (Theodoulidis and Kotouzas 2000) in 2000.

These simulations are valid for the air-core coil based conductivity measurements with "infinite" size of metal plates, consisting of one or several metal layers. The "infinite" size is typically not very significant constraint for real-life measurements, as the eddy current electro-magnetic fields are decreasing quickly in the media. Only a few millimetres larger metal plate than the sensor's size is typically enough to have reasonably precise results.

By using the complex value (real and imaginary parts) measurements of the coil impedance and the inverse model against the previously described simulation modeling, both conductivity value and liftoff between a measurement coil and a measured metal piece could be found. Such solution has been also described in the patent (Snyder 1995).

2 MEASUREMENT AND THEIR INVESTIGATION

For investigation and concept validation, two measurement coils has been prepared:

1. a flat wound coil N14 (inner diameter $d = 3$ mm, outer diameter $D = 20$ mm, $N = 200$ turns, $r_0 = 22.856$ Ω), see figure 1.
2. a PCB based coil array RR1 of 5 coils (for each $N = 40$, outer diameter $D = 14$ mm, $r_0 = 7.441$ Ω), see figure 2.

Impedance of the both types of coils, the N14 and the third (middle) coil RR1-3 of the array RR1 has been measured in the air against 4 metal plates having different conductivities (Fig. 3):

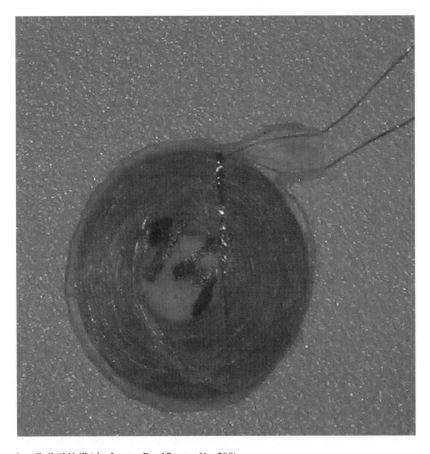

Figure 1.   Coil "N14" ($d = 3$ mm, $D = 17$ mm, $N = 200$).

48

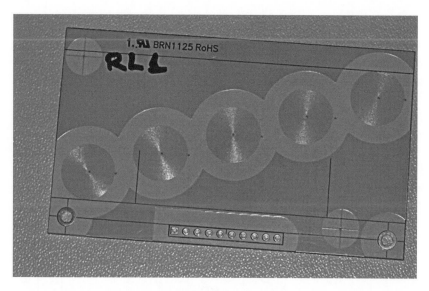

Figure 2.   Planar (PCB) sensor array "RR1" of 5 coils ($N = 40$, $D = 14$ mm), used in the experiments.

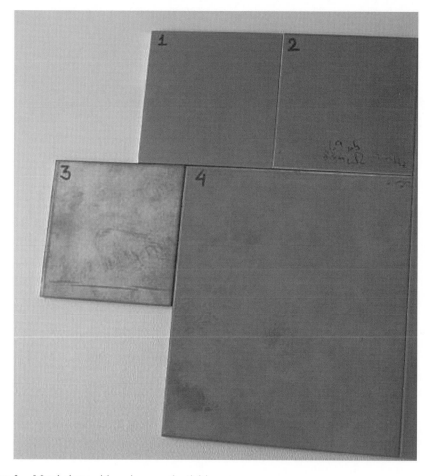

Figure 3.   Metal plates with various conductivities.

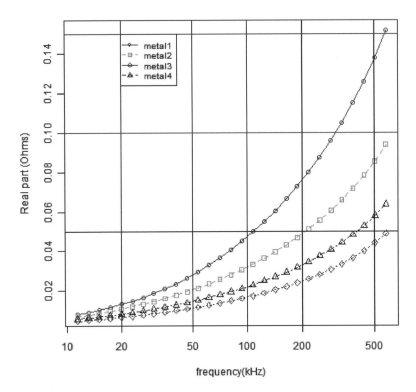

Figure 4. Real part of the N14 coil impedance.

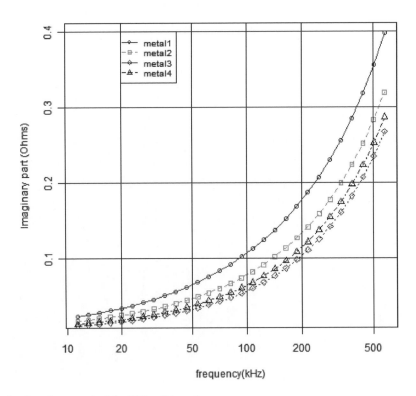

Figure 5. Imaginary part of the N14 coil impedance.

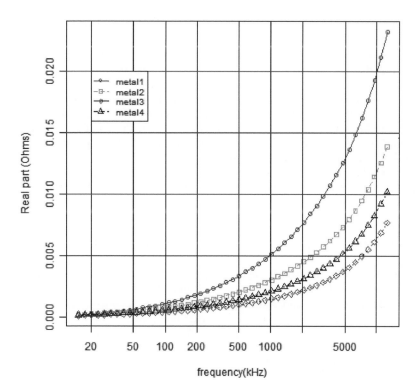

Figure 6.   Real part of the RR1-3 coil impedance.

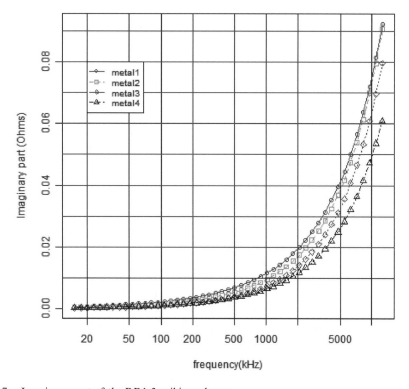

Figure 7.   Imaginary part of the RR1-3 coil impedence.

"metal1" (CuNi, 3.11 MS/m), "metal2" (Nordic Gold, 9,6 MS/m), "metal3" (45 MS/m) and "metal4" ("Brass63", 26.6 MS/m).

As previously described theoretical models consider zero-level real component of the impedance of the coil "in air" (with metal plate or coin near the coil), the ohmic resistance of the coils ($r_0 = 22.585$ Ω for N14 and $r_0 = 7.441$ Ω Ohm for RR1-3) is subtracted from the measured results before their using in the models and showing in the Figures below.

In the experiments approximate lift-off of 0.2 mm has been hold. The following impedance analysers have been used in the experiments: 1) 6500 Series Precision Impedance Analyzer, 20 Hz–120 MHz, from Wayne Kerr Electronics, and 2) HF2IS Impedance Spectroscope, 50 MHz, 210 MSa/s, from Zurich Instruments.

The measured impedances of the measurement coils (in Ω) are given as follows—for N14, the real part in Figure 4 and imaginary part in Figure 5, for RR1-3, the real part in Figure 6 and imaginary part in Figure 7.

3 REVERSE ESTIMATION OF CONDUCTIVITY FROM THE MODEL

To estimate the possibilities for precise measurement of the conductivity by using previously described models of Dodd and Deeds (Dood and Deeds 1968, Dodd and Deed 1975) and Theodoulidis (Theodoulidis and Kotouzas 2000), a reverse simulation model have been developed as a C/C++ software for iterative finding of the best match for lift-off and conductivity of the metal under test. The results of the estimated (from the previously measured real and imaginary parts of the coil impedance) conductivities for metal plates 1 to 4 are given in figures below in the frequency range up to 500 kHz for N14 and up to (?, 10) for RR1-3.

This combined theoretical-experimental investigation shows the possibilities to use eddy current approach with a coil-in-air sensor for reasonably precise (in some % accuracy)

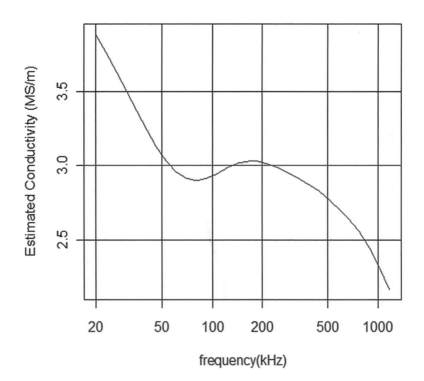

Figure 8.   Back estimated conductivity for N14 coil and "metal1".

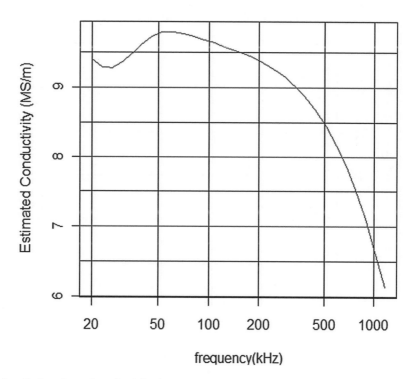

Figure 9.   Back estimated conductivity for N14 coil and "metal2".

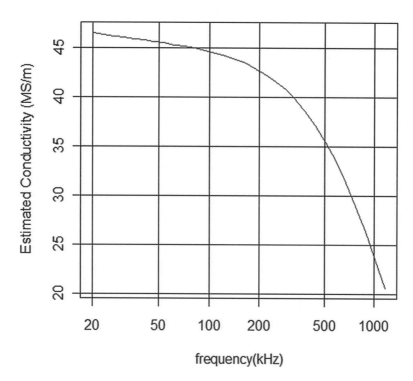

Figure 10.   Back estimated conductivity for N14 coil and "metal3".

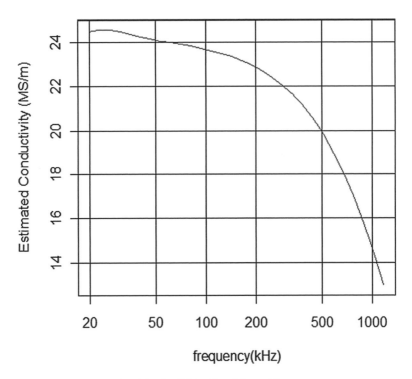

Figure 11.   Back estimated conductivity for N14 coil and "metal4".

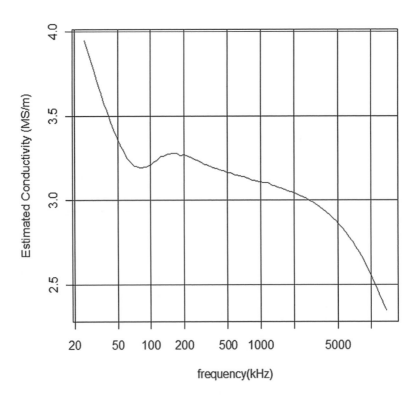

Figure 12.   Back estimated conductivity for RR1-3 coil and "metal1".

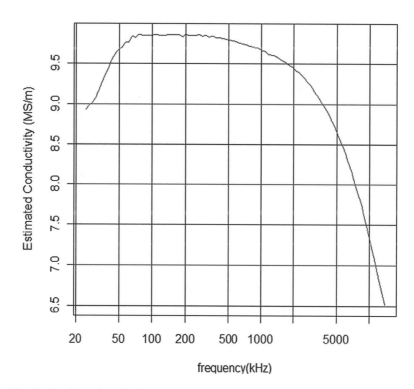

Figure 13. Back estimated conductivity for RR1-3 coil and "metal2".

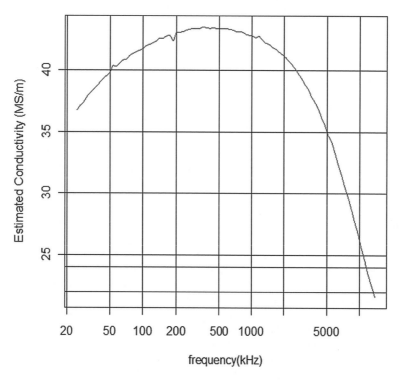

Figure 14. Back estimated conductivity for RR1-3 coil and "metal3".

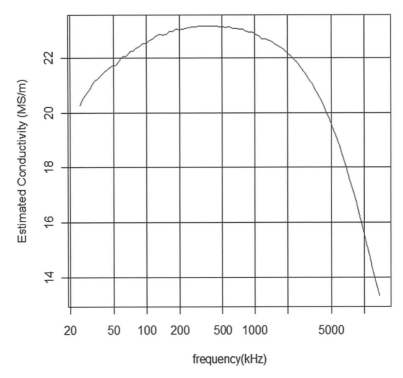

Figure 15.   Back estimated conductivity for RR1-3 coil and "metal4".

electrical conductivity measurement. N14 can be used up to 500 kHz and RR1-3 up to 10 MHz measurements. At higher frequencies, corrections should be introduced to correct systematic frequency response errors.

4   FEM-SIMULATIONS

To validate the used direct- and reverse models, additionally to fitting the experimental data to theoretical (Dodd-Deeds etc) models, also a FEM simulation of the measurement setup (coil above the metal plate) has been done for RR1-3 coil above "metal2" (Nordic gold). COMSOL multiphysics-FEM simulation software package has been sued for such simulation. Got by COMSOL impedance results were used as input to the reverse model of Dodd-Deeds etc, and the results (Fig. 16) show the good correspondence of the models.

5   EXPERIMENTAL MEASUREMENT OF COINS

As PCB based coil array is more reasonable technologically and also has wider frequency range, so the RR1-3 coil was further investigated to be used for coin validation. So, the impedance (real and imaginary parts) against various Euro-coins (5 cent, 50 cents and 1 Euro and 2 Euros) has been measured (at frequencies up to 500 kHz and up to 10 MHz) (Fig. 17–21) and also the "effective" conductivity has been estimated back, according to Dodd-Deeds models, as described before (Figs. 21–24).

   The investigation shows the benefits of using of various (at least two different) frequencies, to distinguish various coins. One possibility (according to Fig. 17–20) seems to be using of one frequency about of 20–30 kHz or another of 100 or 200 kHz. Also adding of a frequency of eg. 5 MHz could be useful.

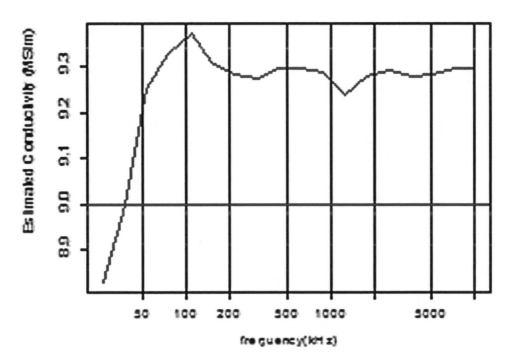

Figure 16. Back estimated conductivity from COMSOL multiphysics..

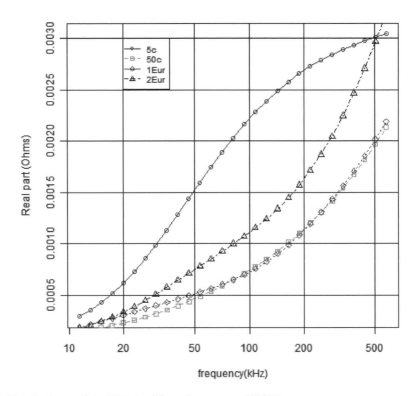

Figure 17. Real part of the RR1-3 coil impedance, up to 500 kHz.

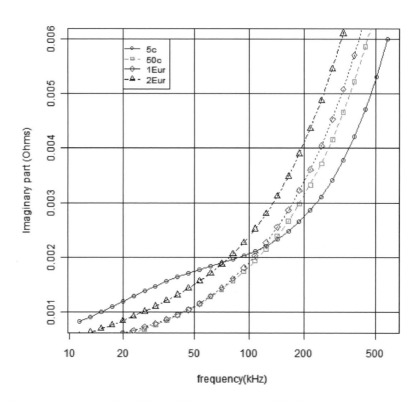

Figure 18.   Imaginary part of the RR1-3 coil impedance, up to 500 kHz.

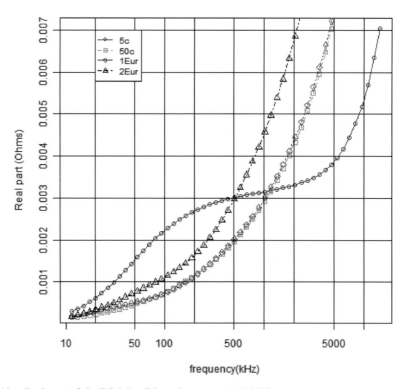

Figure 19.   Real part of the RR1-3 coil impedance, up to 10 MHz.

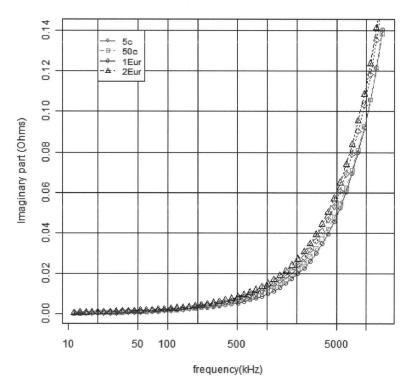

Figure 20.   Imaginary part of the RR1-3, up to 10 MHz.

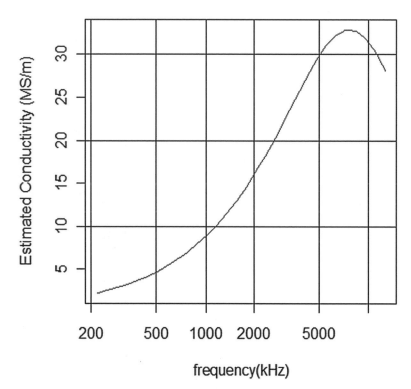

Figure 21.   Back estimated conductivity for 5 cent coin.

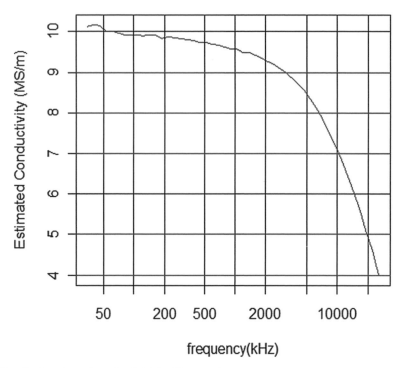

Figure 22.   Back estimated conductivity for 50 cent coin.

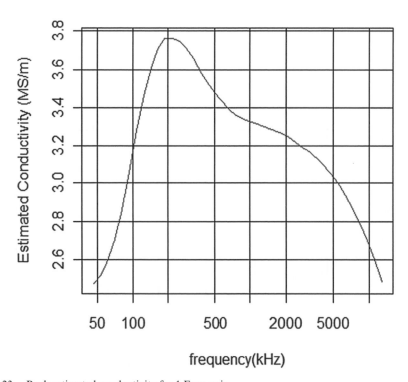

Figure 23.   Back estimated conductivity for 1 Euro coin.

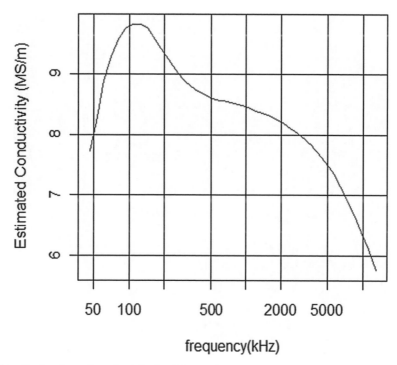

Figure 24.   Back estimated conductivity for 2 Euro coin.

6   COIN VALIDATION SOLUTION

A specially designed test-equipment for having dynamical (in time) test-signals, more-or-less similar to the situation in real coin-validation devices, has been developed (Fig. 25).

Example of the measured signal, for a 20-cent coin, in the time, at 100 kHz frequency, with N14 coil, is given on figure 26 (upper line is a module of the impedance, being similar to the next- imaginary (inductive) component of the impedance and the lowest line is the real part of the coil impedance (change of the losses). The ohmic resistance of the coil is subtracted in the processing, as models deal with ideal coils.

The results show good correspondence of the measured conductivity. For example, in the described experiments, with 20 cent coin (nominal value of the conductivity is about 9.6 MS/m) in the frequency range 40 to 200 KHz, the absolute measurement error is within ±10%.

7   RESULTS, CONCLUSION AND DISCUSSION

So, as described, by using of air-core coil-based sensors (arrays of sensors), relatively high speed and precise conductivity measurement and corresponding coin-validation could be achieved, limited mainly by signal processing solution performance and smart signal approaches used to reduce the needed computational power (and energy). Relaxing the require-ments to needed processing power could be achieved by using synchronous under-sampling, as proposed and implemented in (Märtens and Min 2004, Martens, Land, Gavrijaseva, and Molder 2011). Further improvements can be achieved by using of the smart signal process-ing, for example by using the changeable number of frequencies and sample rates, depending on the position of the coin from sensors (that means from previous measurement results). Further investigation of possibilities of smart signal processing is probably very promising in making the high-speed precise coin-validation more practical.

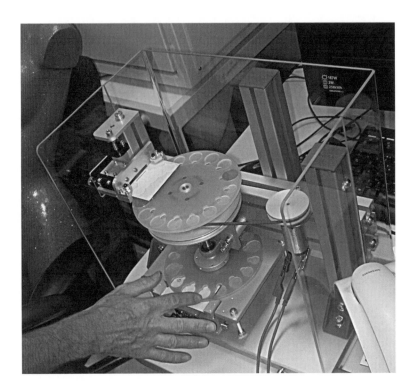

Figure 25.   Auxiliary test-bench for experimental signals.

2cm_coil3_50cent_100kHz

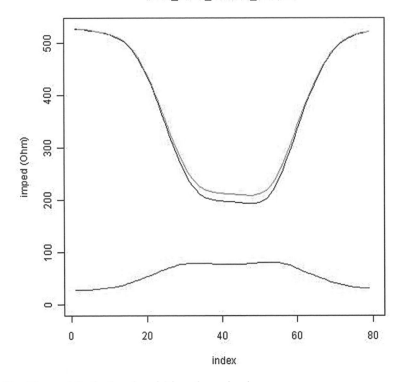

Figure 26.   Measured (in the time domain) impedance signal.

ACKNOWLEDGMENTS

Current work have been supported by EU (FP7-SME project "Safemetal"), Enterprise Estonia (support of Competence Centre ELIKO), target financing SF0142737s06 and grant ETF8905 (Estonian Science Foundation) and by the European Union through the European Regional Development Fund.

Special thanks also to Dr. Theodoros P. Theodoulidis (University of Western Macedonia, Greece) for support with eddy current models.

REFERENCES

Bowler, N. and Y. Huang (2005). Electrical conductivity measurement of metal plates using broadband eddy-current and four-point methods. *Measurement Science and Technology 16*, 2193.

Carlosena, A., A. Lopez-Martin, F. Arizti, A. Martínez-de Guerenu, J. Pina-Insausti, and J. García-Sayés (2007). Sensing in coin discriminators. In *Sensors Applications Symposium, 2007. SAS'07. IEEE*, pp. 1–6. IEEE.

Dodd, C. and W. Deed (1975). Calculation of magnetic fields from time-varying currents in the presence of conductors. Technical report, OAK Ridge National Laboratory.

Dodd, C. and W. Deeds (1968). Analytical solutions to eddy-current probe-coil problems. *Journal of Applied Physics 39*(6), 2829–2838.

Harris, J., J. Churchman, D. Sharman, et al. (2007, July 17). Coin-validation arrangement. US Patent 7,243,772.

Howells, G. (2009, September 8). Coin discriminators. US Patent 7,584,833.

Martens, O., R. Land, A. Gavrijaseva, and A. Molder (2011). Adaptive-rate inductive impedance based coin validation. In *IEEE 7th International Symposium on Intelligent Signal Processing (WISP)*, pp. 1–4. IEEE.

Märtens, O. and M. Min (2004). Multifrequency bio-impedance measurement: undersampling approach. In *Proc. 6th Nordic Signal Processing Symposium. NORSIG*, Volume 2004, pp. 145–148.

Rossiter, P. (1991). *The electrical resistivity of metals and alloys*. Cambridge Univ Pr.

Snyder, P. (1995, February 28). Method and apparatus for reducing errors in eddy-current conductivity measurements due to lift-off by interpolating between a plurality of reference conductivity measurements. US Patent 5,394,084.

Theodoulidis, T. and M. Kotouzas (2000). Eddy current testing simulation on a personal computer. In *Proc. 15th World Conference on Nondestructive Testing*.

Lecture Notes on Impedance Spectroscopy, Volume 3 – Kanoun (ed)
© *2012 Taylor & Francis Group, London, ISBN 978-0-415-64430-3*

Investigation of infection defense using impedance spectroscopy

Anna Schröter & Janette Kothe
Faculty of Electrical and Computer Engineering, Solid-State Electronics Laboratory,
Technische Universität Dresden, Germany

Anja Walther
Department of Paediatrics, University Hospital Carl Gustav Carus, Dresden, Germany
Centre for Translational Bone, Joint and Soft Tissue Research, University Hospital Carl Gustav Carus,
Dresden, Germany

Kerstin Fritzsche & Angela Rösen-Wolff
Department of Paediatrics, University Hospital Carl Gustav Carus, Dresden, Germany

Gerald Gerlach
Faculty of Electrical and Computer Engineering, Solid-State Electronics Laboratory,
Technische Universität Dresden, Germany

ABSTRACT: Neutrophil granulocytes play an important role in the human immune defense. One of their weapons against pathogens is the release of Neutrophil Extracellular Traps (NETs) which can form a biofilm. Biofilms can be detected by electrical impedance spectroscopy (EIS) and enables researchers to observe NET-formation, determine the reaction parameters and to investigate diseases. In this work we present how we adapted the measurement technique to this biological phenomenon and how we obtained first signs of a time-dependent behaviour for this reaction. We found significant changes in the impedance spectra depending on the stimulus. Different stimuli were also part of the study.

Keywords: neutrophil granulocytes, impedance spectroscopy, interdigitated electrodes, neutrophil extracellular traps, biological cells

1 INTRODUCTION

Studying defense mechanisms is a current issue in human biology (Baker, Imade, Molta, Tawde, Pam, Obadofin, Sagay, Egah, Iya, Afolabi, et al. 2008, Clark, Ma, Tavener, McDonald, Goodarzi, Kelly, Patel, Chakrabarti, McAvoy, Sinclair, et al. 2007, Gupta, Hasler, Holzgreve, Gebhardt, and Hahn 2005). Brinkmann et al. (Brinkmann, Reichard, Goosmann, Fauler, Uhlemann, Weiss, Weinrauch, and Zychlinsky 2004, Brinkmann and Zychlinsky 2007) described the formation of Neutrophil Extracellular Traps (NETs) in 2004. After pathogen contact, a lot of neutrophil granulocytes migrate to the source of infection. Their cell core dissolves and the double-stranded DNA meshes with the cell interior containing granula proteins. This bactericide mixture is released through the disrupted cell membrane and spreads within the infection site. The filamentous NET-structures are able to capture the bacteria and fungi and to kill them. Main contents of these NET matrices are chromatin containing mostly double-stranded DNA and histones as well as neutrophil elastase. Figure 1 is a micrograph of NET-structures formed in our laboratory *in vitro*.

Although a lot of information is available about the contents and the release sources, the dynamics of NET-formation is partly unexplored. Electrical impedance spectroscopy can be adapted to the cell environment to facilitate biological studies. Impedancespectroscopy has

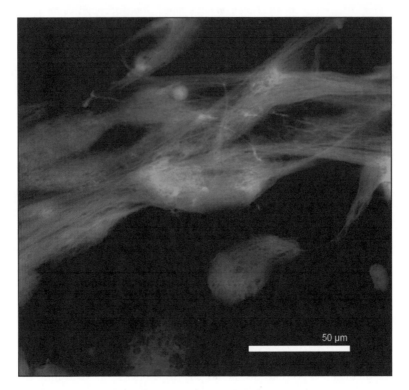

Figure 1.   NETs formed *in vitro* (stained with Sytox Green, 600x, by K. Fritzsche).

arisen as an important tool to observe cell morphology (Brischwein, Grothe, Otto, Stepper, Weyh, and Wolf 2004, Ehret, Baumann, Brischwein, Schwinde, Stegbauer, and Wolf 1997, Ressler, Grothe, Motrescu, and Wolf 2004). Reports range from observation of changes in motility, rearrangements of the cytoskeleton, determination of cell density, proliferation and migration as well as cell adhesion, spreading and membrane effects. Due to the fact that NET-formation is accompanied by similar effects on the cell and changes the medium characteristics substantially, it is useful to use impedance spectroscopy for characterisation of NET-formation dynamics.

2   IN VITRO MEASUREMENTS OF NETS

2.1   *Measurement setup*

As a model environment we used human granulocytes grown on commercial interdigitated electrodes (Roche xCelligence®) in cell medium. The electrodes were coated with poly-L-lysine to ensure adherence of the cells. They are isolated from humanblood and cultivated over night. The cell array is connected to the impedance analyzer with an adapter. The chemical stimulant PMA (phorbol 12-myristate 13-acetate) excites the cells to eject NETs. At high cell concentrations 2 to $4 \cdot 10^6$ in 200 μl a biofilm forms on the electrodes which can be measured with an impedance analyzer (Sciospec ISX-3). The schematic in figure 2 shows the principle of detection.

2.2   *Preliminary investigations*

Preliminary measurements were performed without temperature control. The samples were cultivated in an incubator at 377°C. Afterwards the measurement was carried out at room temperature. This temperature stress does not affect the NET formation but influences the accuracy of impedance determination. Especially at lower frequencies the absolute impedance changes by 24% (Fig. 3). In contrary, the exchange of the culture medium with a stimulation medium results in a small deviation of 0.3%. This means that the effect of stimulation is insignificant and can be neglected. To improve the accuracy, temperature has to be measured to correct the frequency-dependent impedance offset. A calibration curve is given in figure 3. Based on these findings measurements were carried out at stable temperature conditions.

3   REACTION CHARACTERISTICS OF NET-FORMATION

Stimulating neutrophil granulocytes results in a significant change of impedance as displayed in figure 4. The unstimulated control samples were cultivated in the same arrays under the same conditions and measured simultaneously. The impedance change reaches its maximum on average at frequencies about 17.5 kHz after one to four hours with a maximum change of 28 to 55%. Obviously, this reaction varies in a wide range concerning course and dynamics. Still, the impedance change of stimulated samples remain constantly high over a wide range of experiments compared to unstimulated samples which did not change more than 4.8%. Table 1 summarizes these values.

4   CHANGE OF STIMULATION AGENT

In order to understand in vivo infection conditions and to verify our measurements, we applied a different stimulus. Granulocytes react with NET-formation on every kind of

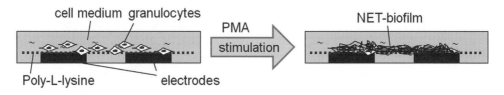

Figure 2.   NET stimulation on electrodes.

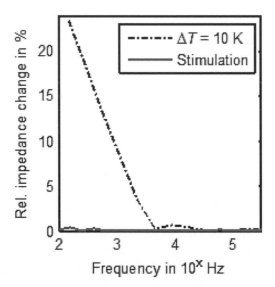

Figure 3.   Relative impedance change due to temperature variability and stimulation.

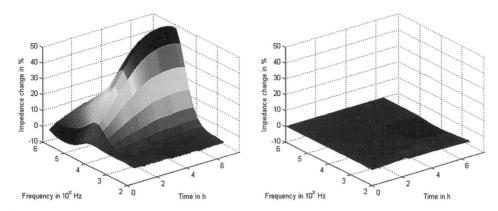

Figure 4.   Impedance change after stimulation (left), unstimulated control sample (right).

Table 1.   Properties of NET-formation.

Property	Value
Time to reach maximum	(1–4) h
Maximum impedance change	(28.55) %
Average frequency at maximum	17.5 kHz
Impedance deviation	
(control sample)	(0.3–0.4)%

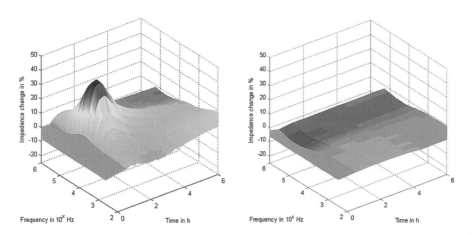

Figure 5.   Impedance change after *S. typhimurium* stimulation (left), unstimulated control sample (right).

pathogen contact, no matter if they are gram-positive and gram-negative bacteria orpar-asites. One interesting bacterium is *Salmonella typhimurium*. Instead of PMA we applied *S. typhimurium* to the granulocyte culture and measured under same conditions. Adding a second cell type could change the impedance spectra itself. So we also watched the deviations caused by the *S. typhimurium* medium.

Figure 5 shows the results for a bacterial stimulation. Compared to unstimulated cell cultures, the stimulated ones undergo a significant change. But the reaction dynamic seems to be different. All stimulated samples reacted very fast (after about one hour) and also dissolved. What is unique is the nearly equal response frequency at 17 kHz. The unstimulated control samples remain stable.

5 CONCLUSION

NET-formation as an immune defense process was investigated using impedance spectroscopy. First measurements demonstrate that EIS is a suitable tool to study dynamics and intensity of the reaction. The structural changes of the cell morphology led to impedance changes even higher than 50%. We saw differences in the reaction with different stimulation agents. Granulocytes stimulated with bacteria reacted similar but with different dynamics. This information could be useful to study differences between different pathogens too. In order to get more information about the correlation between standard NET detection techniques like fluorescence microscopy, a parallel measurement setup has to be established. Afterwards EIS measurement will be useful for investigating the biological defense mechanism of NET-formation with higher validity.

ACKNOWLEDGMENTS

The authors would like to thank the Bundesministerium für Bildung und Forschung (BMBF) for funding the project "ChiBS—Chip-basierte Biologie für die Sensorik" (project number 45952) within the framework of the "WK-Potenzial" programme.

REFERENCES

Baker, V., G. Imade, N. Molta, P. Tawde, S. Pam, M. Obadofin, S. Sagay, D. Egah, D. Iya, B. Afolabi, et al. (2008). Cytokine-associated neutrophil extracellular traps and antinuclear antibodies in plasmodium falciparum infected children under six years of age. *Malar J 7*, 41.

Brinkmann, V., U. Reichard, C. Goosmann, B. Fauler, Y. Uhlemann, D. Weiss, Y. Weinrauch, and A. Zychlinsky (2004). Neutrophil extracellular traps kill bacteria. *Science 303*(5663), 1532.

Brinkmann, V. and A. Zychlinsky (2007). Beneficial suicide: why neutrophils die to make nets. *Nature Reviews Microbiology 5*(8), 577–582.

Brischwein, M., H. Grothe, A. Otto, C. Stepper, T. Weyh, and B. Wolf (2004). Living cells on chip: bioanalytical applications. *Springer Series on Chemical Sensors and Biosensors 2*, 159–180.

Clark, S., A. Ma, S. Tavener, B. McDonald, Z. Goodarzi, M. Kelly, K. Patel, S. Chakrabarti, E. McAvoy, G. Sinclair, et al. (2007). Platelet tlr4 activates neutrophil extracellular traps to ensnare bacteria in septic blood. *Nature medicine 13*(4), 463–469.

Ehret, R., W. Baumann, M. Brischwein, A. Schwinde, K. Stegbauer, and B. Wolf (1997). Monitoring of cellular behaviour by impedance measurements on interdigitated electrode structures. *Biosensors and Bioelectronics 12*(1), 29–41.

Gupta, A., P. Hasler, W. Holzgreve, S. Gebhardt, and S. Hahn (2005). Induction of neutrophil extra-cellular dna lattices by placental microparticles and il-8 and their presence in preeclampsia. *Human immunology 66*(11), 1146–1154.

Ressler, J., H. Grothe, E. Motrescu, and B. Wolf (2004). New concepts for chip-supported multi-well-plates: realization of a 24-well-plate with integrated impedance-sensors for functional cellular screening applications and automated microscope aided cell-based assays. In *Engineering in Medicine and Biology Society, 2004. IEMBS'04. 26th Annual International Conference of the IEEE*, Volume 1, pp. 2074–2077. IEEE.

Lecture Notes on Impedance Spectroscopy, Volume 3 – Kanoun (ed)
© 2012 Taylor & Francis Group, London, ISBN 978-0-415-64430-3

Erythrocyte orientation and lung conductivity analysis with a high temporal resolution FEM model for bioimpedance measurements

Mark Ulbrich
Philips Chair for Medical Information Technology (MedIT), RWTH Aachen University, Aachen, Germany

Jens Mühlsteff
Philips Research, Eindhoven, The Netherlands

Piotr Paluchowski & Steffen Leonhardt
Philips Chair for Medical Information Technology (MedIT), RWTH Aachen University, Aachen, Germany

ABSTRACT: Impedance cardiography (ICG) is a simple and inexpensive method to acquire data on hemodynamic parameters. In this work, the influence of three dynamic physiological sources is analyzed using a model of the human thorax with a high temporal resolution. Simulations are conducted using the finite element method with a temporal resolution of 125 Hz. The ICG signal caused by wave propagation on the aorta shows good agreement with the measured signals (r = 0.75). In addition, it is shown that lung perfusion and erythrocyte orientation have a high impact on the signal, altering the maximum of the curve as well as the characteristic points used to calculate hemodynamic parameters, such as left ventricular ejection time.

Keywords: impedance cardiography, finite element simulations, aortic wave propagation, lung perfusion, erythrocyte orientation

1 INTRODUCTION

Demographic changes in Europe have resulted in an increasingly aging society, which faces increasing numbers of geriatric patients and diseases, leading to increasing costs and burden on medical staff. Therefore, it is important to establish methods to treat elderly persons more (cost-) effectively by improving diagnostics and monitoring of those diseases that are prevalent among the elderly.

A common cause of death in Western Europe is chronic heart failure. Measures for its severity are hemodynamic parameters, including stroke volume (SV). Until now, the gold standard for measuring these parameters is the thermodilution technique using pulmonary artery catheters. However, the risks of estimating cardiac output via catheters include infections, sepsis and arrhythmias, as well as increased morbidity and mortality. An alternative technique to assess SV easily and cost-effectively is the use of impedance cardiography (ICG).

However, ICG is not commonly used as a diagnostic method because it is not yet considered to be valid [Cotter (2006)]. One reason for this is the inaccuracy of the technology itself concerning the calculation of SV. Another reason is that processes in thehuman body during ICG measurements are largely unknown. However, one way to analyze where the electrical current paths run, and which tissue makes a significant contribution to the measurement result, is to use computer simulations employing the finite element method (FEM).

In this paper, the impact of three dynamic sources on the ICG signal is analyzed.

2   BIOIMPEDANCE MEASUREMENTS

For bioimpedance measurements, two outer electrodes are used to inject a small alternating current into the human body and, via two inner electrodes, the voltage drop is measured to calculate the complex impedance. If a frequency spectrum of 5 kHz to 1 MHz is used to measure the bioimpedance for each frequency, this method is called bioimpedance spectroscopy (BIS). This frequency range, called β-dispersion, is generally the most interesting for diagnostic purposes, since physiological and pathophysiological processes lead to changes in body impedances with high dynamics. BIS is commonly used to assess the body composition of humans. If only one frequency of this spectrum is used to measure the bioimpedance continuously, this method is called ICG. Since ICG operates at a certain frequency between 20 and 100 kHz, only one continuous point on a complex Cole curve can be obtained by ICG measurements. Figure 1 shows the measuring principle of ICG.

Here, the measurement technique is implemented with the use of pairwise short-circuited current injecting electrodes. Combined with the fact that the voltage measuring electrodes at neck and abdomen lie on the same equipotential, it is clear that these eight electrodes represent the 4-point measurement technique. By measuring the impedance continuously, time-dependent hemodynamic parameters can be extracted from the measured impedance curve using its temporal derivative (Fig. 2).

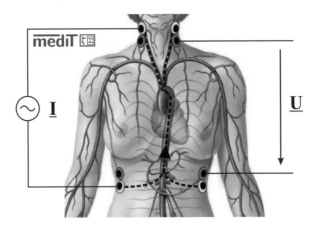

Figure 1.   Measuring principle of impedance cardiography: $\underline{I}$ = complex current, $\underline{U}$ = complex voltage [www.medit.rwth-aachen.de (2010)].

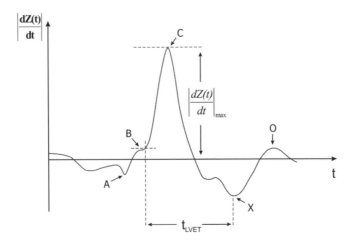

Figure 2.   ICG wave ($|Z(t)|'$) with its characteristic points [Packer (2006)].

The derivative of the continuous impedance $Z(t)$ is the ICG signal whose maximum $|\frac{dZ}{dt}|_{max}$ is used for the calculation of stroke volumes. The SV measured with ICG according to Bernstein and Sramek can be described by the following equation [Van De Water (2003)]:

$$SV = \delta \cdot \frac{(0,17)^3}{4,2} \cdot \left|\frac{dZ}{dt}\right|_{max} \cdot \frac{t_{LVET}}{Z_0} \qquad (1)$$

Here the factor δ is the actual weight divided by the ideal weight, t_{LVET} the left ventricular ejection time and Z_0 the thoracic base impedance.

3 METHODS

Classical ICG analyzes the impedance of the thorax, approximating its volume by one outer cylinder with a conductivity from a mixture of tissues, containing another cylinder representing the aorta. However, this type of assumption leads to modeling errors. Moreover, under certain conditions SV measurements will be unreliable. Therefore, one task is to find alternative techniques and models to improve the reliability of ICG measurements. Initially, this was simulated using the FEM and the Visible Human Dataset from the [National Library of Medicine (1986)]. However, since this dataset has no information on dynamics, a new model was created based on this dataset using simple geometric solids, such as frustums, spheres and cylinders. In addition, this approach reduces simulation time. Figure 3 shows the simplified high temporal resolution (HTR) model whose anatomical geometry is, nevertheless, very accurate.

The model is composed of static volumes from the Visible Human Dataset and new dynamic volumes: aorta, heart, vena cava, carotid vessels, rib cage and lung. For simplicity, ring electrodes are used for current injection at neck and abdomen. It is also possible to use standard spot electrodes for the simulation. Conductivity (σ) and permittivity (ε_r) values for every tissue are implemented using the data according to Gabriel (1996) (Table 1).

The static volumes comprise the tissues fat, muscle and major abdominal organs. The upper surface of these organs is used as a border to create lung tissue as a function of the position of the diaphragm and rib cage, to increase or decrease lung volume according to

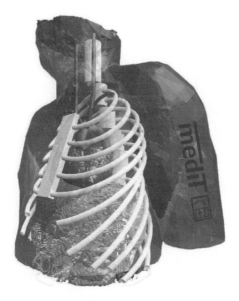

Figure 3. The simulation model.

the respiratory phase. The rib cage itself is constructed using tori, so that respiration can be simulated by altering the angle of these tori around a certain axis. The dynamics of the aorta are implemented based on data derived from a projectin which the arterial system was rebuilt using silicone representing an arterial model [Matthhys (2007)]. In addition, pressure and flow data are acquired at various points on the aorta. Thus, the temporal development of pressure and flow for every point on the aorta can be calculated and interpolated using MATLAB® (MathWorks, Inc. MA, USA). Using these data, the radial change of the aorta can be computed (Fig. 4).

In addition, conductivity changes related to erythrocyte orientation and lung conductivity are analyzed. Due to the alignment of erythrocytes along a laminar flow during high velocities inside blood vessels, blood conductivity changes. Hence, blood conductivity increases during the ejection of blood from the heart. This change in conductivity was taken into account based on experimentally derived data [Raaijmakers (1996)]. Change in the conductivity of lung tissue due to lung perfusion was also simulated using experimentally derived data [Zhao (1996)].

The CST EM Studio® Studio (Computer Simulation Technology, Darmstadt, Germany) was used to perform the simulations. Calculation frequency was set at 100 kHz. Since the

Table 1.   Permittivity and conductivity values at 100 kHz [Gabriel (1996)].

	σ [S/m]	ε_r
Blood	0.70292	5120
Myocard	0.21511	9845.8
Bone	0.020791	227.64
Fat	0.024414	92.885
Muscle	0.36185	8089.2
Abdomen	0.2	4000
Lung	0.10735	2581.3

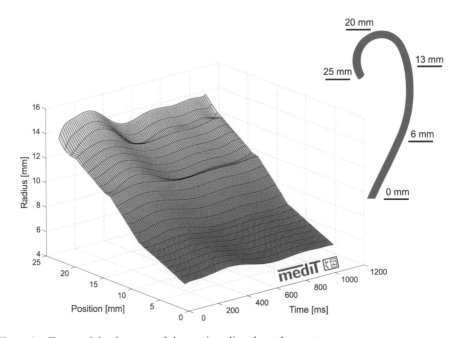

Figure 4.   Temporal development of the aortic radius along the aorta.

74

corresponding wavelength is much higher than our measuring volume, the electroquasistatic solver was used (Eq. 2).

$$\lambda = \frac{c}{f} = \frac{34000 \; \frac{m}{s}}{100 \; \text{kHz}} = 340 \; m \qquad (2)$$

To calculate the complex impedance, a voltage source was used to fix the voltage at the electrode sites. The current was assessed by integrating the conduction current density and the displacement current density over a predefined face, assuming a harmonic oscillation (Eq. 3–5).

$$\underline{I}_{cond} = \int_A (\sigma \cdot \vec{E}) d\vec{A} = \int_A \underline{\vec{J}} d\vec{A} = \int_A (Re\{\underline{\vec{J}}\} + jIm\{\underline{\vec{J}}\}) d\vec{A} \qquad (3)$$

$$\underline{I}_{disp} = \int_A \left(\varepsilon \cdot \frac{\partial \vec{E}}{\partial t} \right) d\vec{A} = \int_A \left(\frac{\partial \underline{\vec{D}}}{\partial t} \right) d\vec{A} = \int_A (\underbrace{j \, \omega Re\{\underline{\vec{D}}\}}_{Im\{\underline{I}_{disp}\}} - \underbrace{\omega Im\{\underline{\vec{D}}\}}_{Re\{\underline{I}_{disp}\}}) d\vec{A} \qquad (4)$$

$$\underline{I}_{total} = \underbrace{Re\{\underline{I}_{cond}\} + Re\{\underline{I}_{disp}\}}_{Re\{\underline{I}_{total}\}} + \underbrace{Im\{\underline{I}_{cond}\} + Im\{\underline{I}_{disp}\}}_{Im\{\underline{I}_{total}\}} \qquad (5)$$

For validation purposes, the simulated curves for 100 kHz were compared to real measured data. These data were acquired using the Niccomo™ device (Medis GmbH, Ilmenau, Germany) on a male subject (weight 75 kg, height 1.8 m).

4 RESULTS

The model consists of all the important tissues with the aorta as volumetric dynamic source, and lung and erythrocyte orientation as conductive dynamic sources (Fig. 3). The dynamic impedance was simulated with a resolution of 125 Hz for all dynamic sources representing ±103 points in time for one heartbeat (Fig. 5).

In general, conductivity changes of lung and erythrocyte orientation have a high impact on the ICG curve compared to the curve resulting from volumetric changes of the aorta. Figure 5 shows that three major influences are involved. First, the maximum of the ICG signal has increased by 3.14 Ω; second, the time between the curve's maximum and the first axis crossing it is prolonged and, third, the results show a shift of the local minimum of the ICG wave (X-point) to an earlier phase of the heart cycle (70 ms).

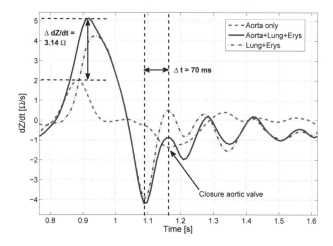

Figure 5.   Simulated ICG signals caused by different physiological sources.

5 CONCLUSION

The aim of this work was to analyze the influence of conductivity changes of lung perfusion and erythrocyte orientation, as well as wave propagation on the aorta, on the ICG signal using FEM simulations.

Several effects were observed. First, simulations with an improved HTR model were conducted using three dynamic sources for the ICG signal. Second, the results of the ICG simulations at 100 kHz for the aortic wave propagation show good agreement with the measured data (r = 0.75). Third, it was shown that conductivity changes have a major impact on the ICG signal and, due to the good correlation of the curve due to volumetric changes of the aorta, other effects are also assumed to alter the signal. One of the influences is a shift of the X-point towards an earlier phase of the heartbeat, thus lowering left ventricular ejection time. This effect might correlate with the results on ICG signals as measured by various researchers, showing that the X-point as an indicator of aortic valve closure is not well defined [Carvalho (2011)]. To substantiate this, more sources have to be included in the simulations.

ACKNOWLEDGMENT

This work was supported by Philips Research Europe and contributes to the project "Heart Cycle" of the European Union.

BIBLIOGRAPHY

Carvalho, P. (2011). Robust characteristic points for ICG: Definition and comparative analysis. *Biosignals—International Conference on Bio-inspired Systems and Signal Processing*.

Cotter, G. (2006). Impedance cardiography revisited. *Physiological Measurement 27*, 817–827.

Gabriel, C. (1996). The dielectric properties of biological tissues. *Physics in Medicine and Biology 41*, 2231–2249.

Matthys, K. (2007). Pulse wave propagation in a model human arterial network: assessment of 1-d numerical simulations against in vitro measurements. *Journal of Biomechanics 40*(15), 3476–3486.

National Library of Medicine (1986). The visible human project. www.nlm.nih.gov/research/visible/visible_human.html

Packer, M. (2006). Utility of impedance cardiography for the identification of short-term risk of clinical decompensation in stable patients with chronic heart failure. *Journal of the American College of Cardiology 47*, 2245–2252.

Raaijmakers, E. (1996). The influence of pulsatile blood flow on blood resistivity in impedance cardiography. *18th International Conference of the IEEE Engineering in Medicine and Biology Society 5*, 1957–1958.

Van De Water, J. (2003). Impedance cardiography—the next vital sign technology? *Chest 123*, 2028–2033.

www.medit.rwth-aachen.de (2010). Homepage Chair for Medical Information Technology. Online.

Zhao, T.-X. (1996). Modelling of cardiac-related changes in lung resistivity measured with EITS. *Physiological Measurement 17*, A227–A234.

Lecture Notes on Impedance Spectroscopy, Volume 3 – Kanoun (ed)
© 2012 Taylor & Francis Group, London, ISBN 978-0-415-64430-3

Influence of meat aging on modified Fricke model's parameter

Mahdi Guermazi

Chair for Measurement and Sensor Technology, Chemnitz University of Technology,
Chemnitz, Germany

Olfa Kanoun

Research Unit on Intelligent Control, Design & Optimization of Complex Systems (ICOS),
Sfax Engineering School, University of Sfax, Sfax, Tunesia

Uwe Tröltzsch

Chair for Measurement and Sensor Technology, Chemnitz University of Technology,
Chemnitz, Germany

Nabil Derbel

Chair for Measurement and Sensor Technology, Chemnitz University of Technology,
Chemnitz, Germany
Research Unit on Intelligent Control, Design & Optimization of Complex Systems (ICOS),
Sfax Engineering School, University of Sfax, Sfax, Tunesia

ABSTRACT: The aim of the investigation is to find the most suitable model that can be used to estimate the electric behavior of cell and biological tissue when measuring the aging of beef and to get results in agreement with physical analyses. For this purpose measurements were done in the frequency range from 40 Hz–10 MHz where two dispersion areas are denoted, dispersion α at low frequency and dispersion β at high frequency. According to literature, the α-dispersion serves in the study of cells membrane integrity during aging. The reliable model that can be used to estimate the electric behavior of cell and biological tissue is the modified Fricke model. This model describes the depressed semicircles and is more suitable to approximate the meat than the Fricke model. The model parameters show that the intercellular resistance increases by aging caused by the shrinkage of fibers and release of electrolytes from cells to the extracellular. Simultaneously this effect results besides cutting of muscle fibers by the enzymes in the decrease of extracellular resistance. This decrease is due also to the inert enzymes that will gradually cut the muscle. The decrease of capacitance in the two cases is due to oxidation of the phospholipids membrane layers that makes the membrane porous.

Keywords: impedance spectroscopy, meat aging, Fricke model, Maxwel Wagner, model parameters, circular electrode

1 INTRODUCTION

Meat characterization can be used in many applications such as control agencies of meat quality, meat industry and domestic use. Different methods, such as ultrasonic, electromagnetic, optical and electrical methods can be principally used for meat aging detection. The electrical method is a promising method due to the possibility of getting much information and realizing measurement systems with low costs and with short measurement times. Biological tissue e.g. meat is an aggregation of cells which can be considered as an ionic conductor, where both intracellular and extracellular fluids are electrolytes because they contain ions which are free to migrate and transport the electrical charge. The total ionic conductivity of a solution depends on the concentration, activity, charge and mobility of all free ions in the solution. The most important ions contributing to the ionic current in biological tissue are K^+, Na and Ca^{2+}, (Grimnes and Martinsen 2008). The aim of the investigation is to find the frequency range of meat aging measurement, to find the most suitable model that can be used to estimate the electric behavior of cell and biological tissue, to measure the aging of beef and to get results in agreement with physical analyses. For this purpose, an experimental investigations were carried out on beef meat of the type rectus abdominus (RA) and Longisimus Dorsi (LD) using circular electrodes during 8 hours, following a logarithmic law and ranging from 40 Hz to 10 MHz. According to literature, two areas of dispersion α and β are denoted in this frequency range. Schwan described these two frequency bands (Schwan 1957). Foster and Schwan (Foster and Schwan 1989) modeled these dispersion types for several biological tissues. Mechanisms at low frequencies like diffusion and lateral movement of ions along cell walls are related to α-dispersion. Current flow is due to the diffusion and lateral movement of the ions along the insulating cell membrane. Almost no current can pass through the cell because of the high resistance of the cell membrane at low frequency. Mechanisms at high frequencies correspond to β-dispersion, which serves in the study of cells membrane integrity during meat aging (Pliquett, Altmann, Pliquett, and Schöberlein 2003). β-dispersion is associated with the dielectric properties of the cell membranes and their interactions with the extra- and intra-cellular electrolytes and it serves in the study of cells membrane integrity during meat aging. One of the phenomena that contribute to the β-dispersion is Maxwell-Wagner effect (Schwan and Foster 1980, Grimnes and Martinsen 2000) due to the interfacial relaxation process occurring in all systems where the electric current must pass an interface between two different materials (Cuela, Barlea, and Barlea 2008). Cell like any particle, will be polarized in response to an electric field. Cellular membrane acts as barriers to the flow of ions between the intra and extracellular fluids. Another contribution to β dispersion is the relaxation caused by protein and amino acids residues (Schwan 1957, Morgan and Green 2003).

One of the first successful electrical models for the biological tissues was introduced by Fricke and Morse (Fricke 1932, Fricke and Morse 1925). It consists of a resistive element R_e

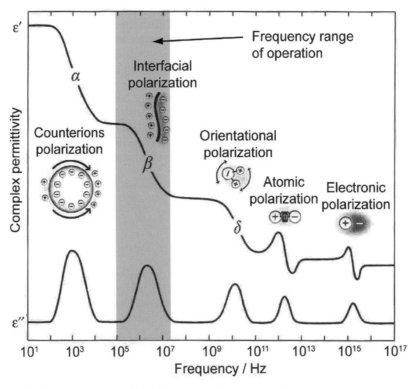

Figure 1.   Mechanisms at low and high frequency, (Demmiere 2008).

representing extracellular fluids placed in parallel with the capacitive element C, representing insulating cell membranes in series with a resistive element R_i representing intracellular fluids. The Fricke model has been extensively used and, even today, some authors make use of it. Damez et al. (Damez, Clerjon, Abouelkaram, and Lepetit 2007) investigated the dielectric behavior of beef meat in the frequency range of 1–1500 KHz. Additionally they worked on the probing of the muscle food anisotropy for meat aging control (Damez, Clerjon, Abouelkaram, and Lepetit 2008). Pliquett et al., investigated the high electrical field effects on cell membranes (Pliquett, Altmann, Pliquett, and Schöberlein 2003). Haemmerich et al. (Haemmerich, Ozkan, Tsai, Staelin, Tungjitkusolmun, Mahvi, and Webster 2002) also used Fricke model during the measurement of changes in electrical resistivity of swine liver after occlusion and postmortem. Konishi et al., used it in electrical properties of extracted rat liver tissue, (Konishi, Morimoto, Kinouchi, Iritani, and Monden 1995). However, it was observed that the Fricke model was not accurate enough to fit the experimental results (Cole and Cole 1941, Cole 1968). By studying the dispersion and adsorption on the dielectric, Cole and Cole introduced the first mathematical expression able to describe the depressed semicircles by finding the behaviour of constant phase element (CPE) that describes a dispersion capacity or change of capacity depending on the frequency. In this work we will measure the electrical impedance of meat of beef to determine the frequency ranges of different dispersion areas. We will analyze the influence of model parameters of modified Fricke model and study them according to the biological basics.

2   EXPERIMENTAL PROCEDURE

The measurements were performed using a probe consisting of circular electrode coated with gold and were recorded on the laboratory impedance analyzer Agilent 4294 A (Fig. 2), following a logarithmic law, ranging from 40 Hz to 10 MHz. A sinusoidal signal was used as excitation signal with a fixed voltage of 0.5 V. Two samples of beef (380.2 Kg, 2 years), Rectus abdominis (RA) and Longisimus Dorsi (LD) were used. The weight of the samples was 112 g at the beginning of experiment that was considered the reference time (24 hours after slaughtering). The measurement was performed four times, the first measurement at

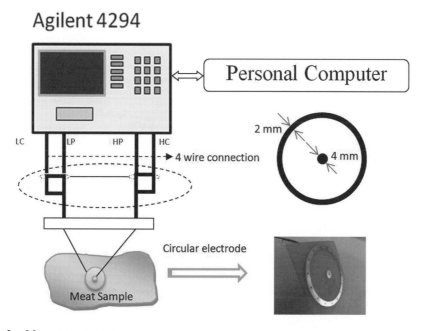

Figure 2.   Measurement setup.

the reference time, time 0 then after 2, 6 and finally 8 hours. The meat was placed in room temperature (23°C); the pH value of the two samples is 6.62 for LD and 5.57 for RA. The fluid loss at the end of experiment for the two samples was 10%.

3   RESULTS AND DISCUSSION

The measurements (Fig. 3) show two areas of dispersion α, β in the frequency range 40 Hz–10 MHz. Measured impedance is analyzed by evaluating characteristic points in the impedance and admittance representation and by fitting several impedance models. Characteristic points (Fig. 3) allow analyzing the impedance evolution over time. The characteristic

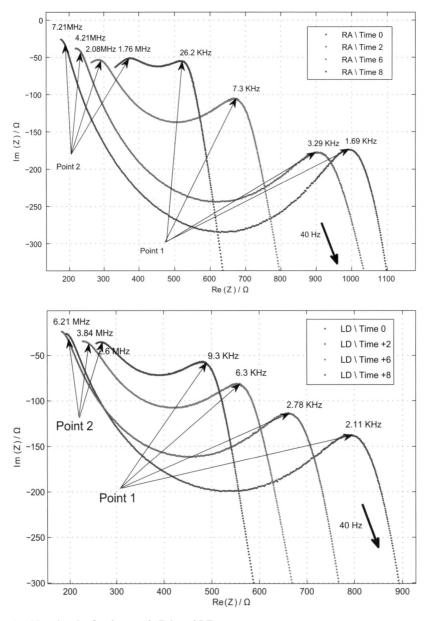

Figure 3.   Nyquist plot for the muscle RA and LD.

80

point 1 is related to the transfer region between α and β dispersion. It is defined at the maximum imaginary part of the impedance or analogous to the minimum imaginary part of the admittance. The characteristic point 2 is defined similar to point 1 but related to the transfer region between α and β dispersion. A depressed semicircle is denoted during the impedance spectrum in the Nyquist plot (imaginary part versus real part), therefore we use the modified Fricke model (McAdams and Jossinet 1995) (Eq. 1), (Fig. 4) in order to consider the CPE-behavior in the spectrum. Modified Fricke model consists of modification of the capacitive element in the Fricke model (Eq. 2), (Fig. 4) by a constant phase element (CPE). Both of them, the capacitive element and the constant phase element are representing insulating cell membranes.

$$Y_{model} = \frac{1}{R_e} + \frac{1}{\frac{1}{k_a(j\omega)^\alpha} + R_i} \tag{1}$$

$$Y_{model} = \frac{1}{R_e} + \frac{1}{\frac{1}{j\omega C} + R_i} \tag{2}$$

The impedance of CPE is described according to eq. 3 (Zoltowski 1998). When α is close to 0, the CPE describes a resistance, close to -1 it describes an inductance, close to 1 it describes a capacity and finally, for the value of 0.5, the result is equivalent to the Warburg diffusion impedance.

$$Z_{CPE} = \frac{1}{k_a(j\omega)^\alpha} \tag{3}$$

Figure 5 shows an example of fitted data to an equivalent circuit using Fricke and modified Fricke model. During the first step of data analysis, data are fitted to an equivalent circuit described by a model equation. For the estimation of the model parameters an evolutionary algorithm is used, described in (Büschel, Tröltzsch, and Kanoun 2011, Kanoun, Tröltzsch, and Tränkler 2006). This algorithm is based on a stochastic global optimization method.

The results of the fitting procedure can be evaluated using residual analysis (Fig. 6) according to equation 5. When comparing the residuals for the Fricke and modified Fricke model it can be seen that the modified Fricke model yields much better results ($\leq 5\%$) compared to Fricke model (up to 20%).

$$R(\omega, \vec{x}, Z_{meas}) = Z(\omega, \vec{x}) - Z_{meas}(\omega) \tag{4}$$

Figure 7 illustrates the modified Fricke model parameters evolution over time, the resistive element R_i representing intracellular fluids increases over time with ΔR_i (LD) = 202.7 Ω and ΔR_i (RA) = 271.5 Ω. Biological tissues are composed about 60% of fluid, two thirds of the fluid is inside the cell and the remaining third in the extracellular (Kanoun, Tröltzsch, and Tränkler 2006).

The fluid loss results in the shrinkage of fibers during the postmortem period and the release of electrolytes from cells to the extracellular tissue (Valet, Silz, Metzger, and

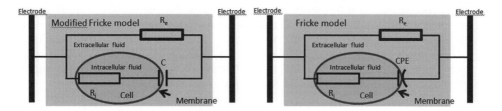

Figure 4. Fricke model and modified Fricke model.

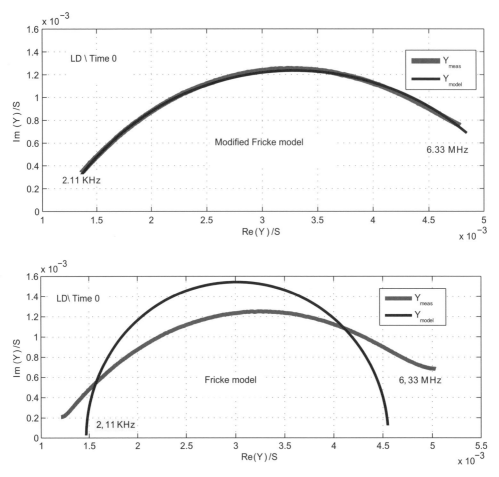

Figure 5. Fitted parameters for the muscle LD at time 0 using Fricke model and the modified Fricke model.

Ruhenstroth-Bauer 1975). This phenomenon causes the increase of R_i. The resistive element Re that represents extracellular fluids decreases over time with ΔR_i (LD) = 331.93 Ω and ΔR_i (RA) = 455.54 Ω. At slaughter, the animal muscles contract due to the release of lactic acid contained in muscle fibers, by releasing the muscle fibers, lactic acid activates the inert enzymes already present in the muscle, which will be gradually cut the muscle fibers. Besides the release of electrolytes from cells to the extracellular tissue, these two phenomena cause the increase of conductivity and proportionally the decrease of R_e.

The fitted ka value does not directly represent the capacitance. Before analysis it needs to be converted into a capacitance by the equations given in Eq. 5 (Hsu and Mansfeld 2001). Where ω' correspond to the frequency corresponding to the maximum value of the imaginary part of impedance in the dispersion area β that corresponds to point 2 (Fig. 4). The capacitance decreases over time with ΔC_{LD} = 0.54972 nF and ΔC_{RA} = 0.87458 nF due to the oxidation of the phospholipids membrane layers that makes the membrane porous. An electrical charging of lipid membranes causes electroporation. This leads to the increase of conductivity and the decrease of capacity (Pliquett, Joshi, Sridhara, and Schoenbach 2007).

$$C = k_a (\omega')^{\alpha - 1} \tag{5}$$

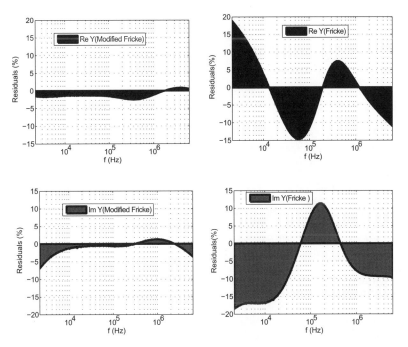

Figure 6.   Residual calculation for the muscle LD at time 0.

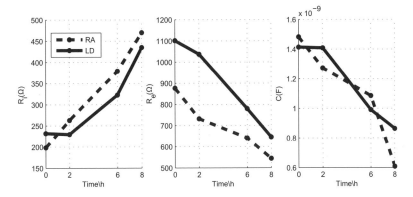

Figure 7.   Parameters evolution over time.

4   CONCLUSION

To summarize the impedance spectrum of meat was modeled according to the modified Fricke model in order to consider the CPE-behavior in the spectrum. The result showed that intercellular resistance R_i increase by aging due to the fluid loses that causes the shrinkage of fibers and release of electrolytes from cells to the extracellular tissue. Simultaneously this effect results, besides cutting of muscle fibers by the enzymes, in the decrease of extracellular resistance R_e. The decrease of capacitance is due to oxidation of the phospholipids membrane layers that makes the membrane porous. This investigation has served to know that circular electrode ensures good measures for short periods of time (8 hours). The circular electrode need to be tested for longer period. β-dispersion could serve in the study of cells membrane integrity during meat aging. Modified Fricke model gives a good representation for the measured data and is more suitable to approximate the meat. We can conclude that experimental results are consistent with the theoretical modelling.

ACKNOWLEDGMENTS

The author kindly appreciates the finacial support of the Deutscher Akademischer Austausch Dienst (DAAD).

REFERENCES

Büschel, P., U. Tröltzsch, and O. Kanoun (2011). Use of stochastic methods for robust parameter extraction from impedance spectra. *Electrochimica Acta 56*, 806–8077.

Cole, K.S. (1968). *Membranes, ions, and impulses: a chapter of classical biophysics*, Volume 1. University of California Press.

Cole, K.S. and R.H. Cole (1941). Dispersion and absorption in dielectrics I. Alternating current characteristics. *The Journal of Chemical Physics 9*, 341.

Culea, E., N.M. Barlea, and S.I. Barlea (2008). Maxwell-wagner effect on the human skin. *Romanian J Biophys 18*(1), 87–98.

Damez, J.L., S. Clerjon, S. Abouelkaram, and J. Lepetit (2007). Dielectric behavior of beef meat in the 1–1500 kHz range: simulation with the Fricke/Cole-Cole model. *Meat Science 77*(4), 512–519.

Damez, J.L., S. Clerjon, S. Abouelkaram, and J. Lepetit (2008). Electrical impedance probing of the muscle food anisotropy for meat ageing control. *Food Control 19*(10), 931–939.

Demmiere, N. (2008). *Continuous-Flow Separation of Cells in a Lab-on-a- Chip using "Liquid Electrodes" and Multiple-Frequency Dielectrophoresis*. Ph.D. thesis, Ecole Polytechnique Fédéral de Lausanne.

Foster, K.R. and H.P. Schwan (1989). Dielectric properties of tissues and biological materials: a critical review. *Critical Reviews in Biomedical Engineering 17*, 25–104.

Fricke, H. (1932). Xxxiii: The theory of electrolytic polarization. *Philosophical Magazine Series 7 14:90*(90), 310–318.

Fricke, H. and S. Morse (1925). The electric resistance and capacity of blood for frequencies between 800 and 41/2 million cycles. *The Journal of General Physiology 9*(2), 153–167.

Grimnes, S. and Ø. Martinsen (2000). *Bioimpedance and Bioelectricity Basics*. Academic Press.

Grimnes, S. and Ø. Martinsen (2008). *Bioimpedance and Bioelectricity Basics*. Academic press.

Haemmerich, D., O.R. Ozkan, J.Z. Tsai, S.T. Staelin, S. Tungjitkusolmun, D.M. Mahvi, and J.G. Webster (2002). Changes in electrical resistivity of swine liver after occlusion and postmortem. *Medical and Biological Engineering and Computing 40*(1), 29–33.

Hsu, C. and F. Mansfeld (2001). Technical note: concerning the conversion of the constant phase element parameter y0 into a capacitance. *Corrosion 57*(09), 747–748.

Kanoun, O., U. Tröltzsch, and H.R. Tränkler (2006). Benefits of evolutionary strategy in modeling of impedance spectra. *Electrochimica acta 51*(8–9), 1453–1461.

Konishi, Y., T. Morimoto, Y. Kinouchi, T. Iritani, and Y. Monden (1995). Electrical properties of extracted rat liver tissue. *Research in Experimental Medicine 195*(1), 183–192.

McAdams, E.T. and J. Jossinet (1995). Tissue impedance: a historical overview. *Physiological measurement 16*, A1.

Morgan, H. and N.G. Green (2003). *AC Electrokinetics: Colloids and Nanoparticles*. Research Studies Press, Ltd.

Pliquett, U., M. Altmann, F. Pliquett, and L. Schöberlein (2003). Py—a parameter for meat quality. *Meat science 65*(4), 1429–1437.

Pliquett, U., R.P. Joshi, V. Sridhara, and K.H. Schoenbach (2007). High electrical field effects on cell membranes. *Bioelectrochemistry 70*(2), 275–282.

Schwan, H. (1957). Electrical properties of tissue and cell suspensions. *Advances in Biological and Medical Physics 5*, 147–209.

Schwan, H.P. and K.R. Foster (1980). Rf-field interactions with biological systems: electrical properties and biophysical mechanisms. *Proceedings of the IEEE 68*(1), 104–113.

Valet, G., S. Silz, H. Metzger, and G. Ruhenstroth-Bauer (1975). Electrical sizing of liver cell nuclei by the particle beam method. mean volume, volume distribution and electrical resistance. *Acta Hepato-Gastroenterol 22*, 274–281.

Zoltowski, P. (1998). On the electrical capacitance of interfaces exhibiting constant phase element behaviour. *Journal of Electroanalytical Chemistry 443*(1), 149–154.

Lecture Notes on Impedance Spectroscopy, Volume 3 – Kanoun (ed)
© *2012 Taylor & Francis Group, London, ISBN 978-0-415-64430-3*

Comparison between state of charge determination methods for lithium-ion batteries based on the measurement of impedance, voltage and coulomb counting

Georg Fauser & Mareike Schneider
Fraunhofer-Institut für Keramische Technologien und Systeme, IKTS, Dresden, Germany

ABSTRACT: Goal of this work was to compare different methods of determining the state of charge (SoC) of lithium-ion-batteries and to investigate the application of these methods on different types of cells. To evaluate the SoC determination techniques, voltage and impedance data were recorded and referenced to the current integrated over time (coulomb counting) and to cell temperature. For 100 randomly spread test measurements for each cell, the SoC was calculated and the difference towards the value indicated by coulomb counting was noted. Regarding the $Li(NiCoMn)_{1/3}O_2$-type cell, determining the SoC by using the open-circuit voltage gave the best results. For the $LiFePO_4$-type cell the most accurate SoC values were achieved by measuring the voltage under load. Impedance based SoC determination gave less precise results. This is mainly due to the strong temperature dependency of the internal resistance. Also not for both cells under test a unique correlation between impedance and SoC could befound.

Keywords: State of Charge, SoC, Impedance, Open Circuit Voltage OCV, Voltage under Load, Coulomb Counting

1 INTRODUCTION

Accurate and reliable state of charge (SoC) determination is a key parameter for operating the Battery Management System (BMS) of Li-ion batteries. Especially in electric vehicles this is a safety-critical issue. The state-of-the-art of SoC determination is integrating the battery current over time (coulomb counting) and/or comparing the measured cell voltage to a predetermined voltage-SoC characteristic. Former is known to give precise results, but parasitic errors are adding up over time and in the caseof loss of the stored value of the integrated current a full charge/discharge of the battery would be necessary to re-establish the SoC calculation (Schalkwijk and Scrosati 2002). The latter is mainly used to gauge the SoC in consumer electronics. Despite its advantage of being able of determining the SoC in any operation condition, it is assumed that it cannot be utilised on cells which show a flat voltage plateau like those based on lithium iron phosphate or lithium titanate (Thomas-alyea 2009).

The impedance of Li-ion cells is known to be dependent on the SoC (Schalkwijk and Scrosati 2002, Tröltzsch 2005). Therefore fixed-frequency impedance (FFI) measurements were carried out and used to calculate the SoC.

Its accuracy was then compared with the SoC calculated with data from voltage under load (VUL), open circuit voltage (OCV) and fixed frequency impedance (FFI).

2 EXPERIMENTAL

All tests were performed on a 10 Ah cylindric cell based on LiFePO4 / graphite (LFP), and on a 10 Ah prismatic cell based on $Li(NiCoMn)_{1/3}O_2$/graphite (NCM). Measurements were carried out on Scinelab cell testers coupled to temperature test chambers from CTS. The experiment was structured in two stages: Part A was building up a database of voltage and impedance data referenced to SoC and cell temperature. Part B contained 100 randomly spread test measurements to evaluate the accuracy of the SoC determination.

2.1 Database build up

The cells were fully charged, then discharged in steps of 0.5 Ah. At the end of each step the VUL was recorded. After cutting off the current a pause of 30 min with 0 A was held. Thereafter the OCV was recorded and FFI was measured at 1 kHz, 10 Hz, 1 Hz and 0.1 Hz. When reaching the discharge cut-off voltage of the cell, the whole test was repeated with a changed environmental temperature. Those were altered in 5 K steps between 0°C and 40°C. The temperature and the depth of discharge (DoD) in Ah was recorded with each measured value. From the DoD (in Ah) the residual charge (Q_r in Ah) was calculated as:

$$Q_r = Q_{max,\vartheta} - \text{DOD} \tag{1}$$

Figure 3 visualises the database for the 0.1 Hz impedance of the NCM cell.

2.2 SOC determination

Subsequently the measurements described in step A were repeated with randomised values. Therefore 10 discharge steps each for 10 temperatures were selected so that they filled the test field between 0 and 40°C resp. 0 and 100% SoC. All temperature and SoC values were then randomised about 5% around their value. Based on that, the test bench set temperature and SoC (by controlling the current over time) for the cell under test and started the measurements. With the obtained values the residual charge could be interpolated out of the look up tables as visualised in Figure 3. Comparing this interpolated value with the residual charge determined by coulomb counting gives the SoC-error (in%) as:

$$\text{SoC}_{error} = \text{SoC}_{measured} - \text{SoC}_{Coulomb-counting} \tag{2}$$

The results are represented as error matrices to analyse the causal relation between SoC-error, state of charge and temperature (Fauser 2010).

3 RESULTS AND DISCUSSION

The SoC-error matrices are shown in Figure 1, where the error is represented as colour. For interpreting the errors the matching LuTs are shown in Figure 3.

Calculating the SoC by measuring the voltage under load (Fig. 1a) and b)) resulted in surprisingly accurate results for the LFP cell. For low SoC the errors are about 2%, for high SoC 5%. The increased deviation is caused by the flat discharge characteristic. Errors for the NCM cell were also about 2% except for an increased error at low temperatures of unknown origin. The advantage of the method is that measuring the voltage with an error of about 2 mV, as in the conducted tests, is technically simple and has to be performed anyway for safety reasons. The main disadvantage is that of course the voltage under load is dependent of the load current. Therefore, when applying this method, a charge/discharge model of the cell must be available. Also this model has to be automatically adapted to the aging of the cell during its lifetime.

The results for the SoC calculated from OCV data are shown in Figure 1 c) and d). As expected calculating the SoC with OCV data resulted in large errors of up to 25% for the LFP cell. The reason for this is its flat voltage plateau at about 50% (Fig. 2 c). With a SoC error of only 2% this method was shown to be applicable for the NCM type. A disadvantage of OCV measurements is the time needed to reach a full relaxation of the cell voltage, which can take from 10 minutes to 2 hours. On the other hand the comparatively very low dependency of the OCV from internal resistance and temperature makes the method very robust.

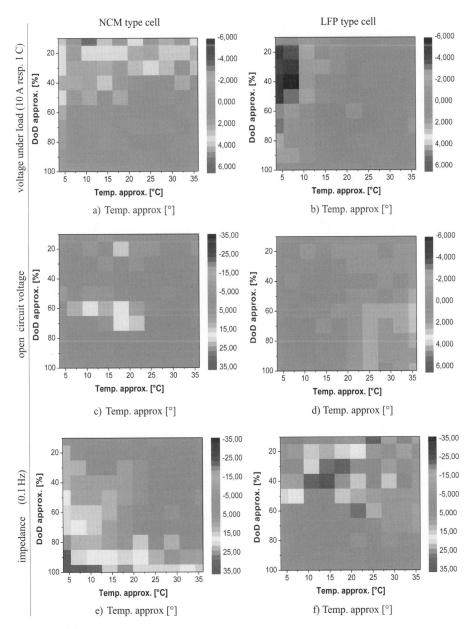

Figure 1. SoC-error matrices for voltage under load (1st row), open circuit voltage (2nd row) and 100 mHz fixed frequency impedance (3rd row). Results for the LFP cell are displayed in the left, NCM in the right column. Scales show the absolute deviation between calculated SoC and that determined by coulomb counting in%.

Impedance measurements showed, especially at low frequencies, a correlation with the SoC. However the SoC errors shown in Figure 1 e) and f) are much larger than those from the voltage measurements. For the LFP cell the errors increase with depth of discharge and decreasing temperature up to 35%. In the corresponding LuT (Fig. 2e) it is visible that in this area the space between the characteristic lines is too large to interpolate the residual charge within an acceptable margin. Blanking out the impedance-peaks at 100% SoC, which are visible in the LuT, gave good results of about 5% error for high SoC. For the NCM cell errors of 30% were calculated in the range of high SoC. Those are due to the partial

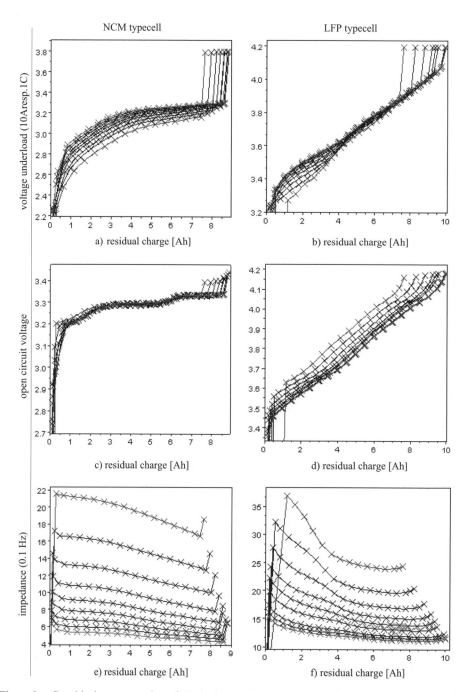

Figure 2. Graphical representation of the look up tables which are the base for the SoC calculations (corresponding to the layout of Figure 1).

independence of the 100 mHz impedance from SoC (Fig. 2f). The impedance measuring time was 100 s since ten periods of the signal were recorded. The calculation effort is here much higher than for a simple voltage measurement, but the more important drawback is the costly hardware needed. This can be reduced when instead of a sine signal an impulse or the load current from e.g. an electric car is used as excitation. This method is theoretically possible but not yet implemented in commercially available measurement equipment or battery

88

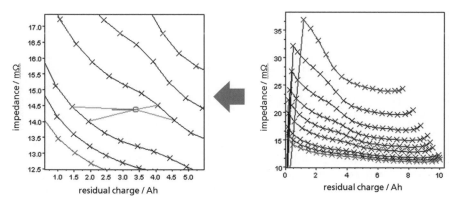

Figure 3.   Test.

Table 1.   Comparison between different methods of determining the SoC of lithium-ion batteries.

		Voltage				Impedance				Coulomb counting
		Under load		OCV		10 Hz		0.1 Hz		
Criteria	Method	LFP	NCM	LFP	NCM	LFP	NCM	LFP	NCM	
Performance	deviation max[A5]	5%	2%	25%	6%	x	>30%	35%	30%	–
	expected[A6]				2%		10%	<10%	10%	
	robustness	good		good[A1]		fair[A2]				fair[A3]
Effort	hardware	low				high				fair[A7]
	data analysis	simple				complex				Simple
	measuring time	instant[A4]		5–60 min		ca. 1 s		>1 min		–

[A1] errors due to aging and temperature dependency of the internal resistance are less grievous than for measuring voltage under load
[A2] method is very sensitive to measurement errors
[A3] parasitic errors are adding up over time and in the case of loss of the stored value of the integrated current no estimation of the SoC is possible
[A4] further experiments need to clarify whether a constant current has to flow through the cell for a certain time to obtain a reliable measurement
[A5] maximum SoC error within the test field
[A6] it is assumed that by removal of algorithmic and measurement errors the accuracy of the SoC determination can be further improved
[A7] a current sensor with a high accuracy is needed

management systems. Impedance measurements are regarded as a convenient method for determining the degradation of the cell due to aging (state of health, SoH). Therefore it is still likely that impedance spectroscopy will in future be implemented in BMS for large scale Li-ion batteries.

Since measurements of impedances at higher frequencies resulted in SoC-errors of up to 50%, those results are not shown here.

4   CONCLUSION

The results of the SoC tests are summarised in table 1. In this test the SoC calculated based on 100 mHz impedance measurements showed less accurate results than that based on voltage measurement. Impedance measurements need a comparatively high effort in hardware and calculation time. Therefore its application is only assumed to be sensible when impedance data is collected anyway for cell diagnosis.

Especially for the shown impedance measurements the used simple look up table method was not able to cope with the strong temperature dependency of the measurements. Nevertheless, the accuracy of the SoC determination might be improved by applying a more sophisticated method of deriving the SoC information from the database, e.g. a model based approach.

Coulomb counting was used as reference in the experiments presented here. Despite its initially mentioned disadvantages it is likely to remain state of the art in SoC determination. The investigations confirmed the point that the best way to counteract the drawbacks of the individual methods and thereby to achieve the best results would be to combine different methods e.g. by a fuzzy-logic approach (Schalkwijk and Scrosati 2002). It was also shown that the SoC determination has to be optimised for each type of Li-ion chemistry since they differ considerably in their characteristics.

ACKNOWLEDGMENTS

This project was supported by funds from *Europäische Fonds für regionale Entwicklung* (*EFRE*) and the *Freistaat Sachsen*.

REFERENCES

Fauser, G. (2010). Charakterisierung und vergleich von lithium-ionen-batterien für den einsatz in elektrofahrzeugen. diplomarbeit. Master's thesis, TU-Dresden/Fraunhofer IKTS.

Schalkwijk, W. and B. Scrosati (2002). Advances in lithium ion batteries introduction. In W. Schalkwijk and B. Scrosati (Eds.), *Advances in Lithium-Ion Batteries*, pp. 1–5. Springer US. 10.1007/0-306-47508-1_1.

Thomas-alyea, K. (2009, June 3). Method and system for determining state of charge of an energy delivery device. US Patent App. 12/477,382.

Tröltzsch, U. (2005). *Modellbasierte Zustandsdiagnose von Gerätebatterien*. Ph.D. thesis, Universität der Bundeswehr München, Universitätsbibliothek.

Lecture Notes on Impedance Spectroscopy, Volume 3 – Kanoun (ed)
© *2012 Taylor & Francis Group, London, ISBN 978-0-415-64430-3*

Microwave-assisted synthesis and properties of nanosized LiMPO$_4$ (M = Mn, Co)

Christoph Neef, Carsten Jähne & Rüdiger Klingeler
Kirchhoff Institute for Physics, University of Heidelberg, Heidelberg, Germany

ABSTRACT: We report on the microwave assisted hydrothermal synthesis of LiMnPO$_4$ and LiCoPO$_4$ and their electrochemical characterization by means of cyclic voltammetry and potentioelectrochemical impedance spectroscopy (PEIS/SPEIS). The influence of various synthesis parameters like pH-value and reaction time on the particle shape and size and thus on the electrochemical performance are studied. In addition, effects of the battery state of charge on the impedance spectra are investigated and discussed.

Keywords: olivine nanostructures, cyclic voltammetry, electrochemical, impedance spectroscopy

1   INTRODUCTION

During the last decade transition metal-based nanomaterials have received a lot of attention because of their wide field of potential applications like sensors, solar cells, and battery materials. From a fundamental point of view, this potential arises not only from the mere size reduction but from fundamentally novel properties appearing upon size reduction which may yield interesting new applications for well-known bulk materials. High surface area and reduced ionic or/and electronic diffusion lengths as well as less tendency to structural changes compared to the bulk form render nanoscaled materials beneficial for electrochemical energy storage. In this respect, olivine nanostructures are particularly promising because in olivine-type lithium iron phosphates the underlying FePO$_4$-network is stable up to high temperatures and offers cycling stability at high cell voltages (Padhi, Nanjundaswamy, and Goodenough 1997, Goodenough 2007). However, ionic and electronic conductivity are relatively low in this material. Detailed studies of the intercalation process show that only the phases LiFePO$_4$ and FePO$_4$ where the Fe-ions exhibit the single oxidation states 2+ and 3+, respectively, are present in the cathodes which are bad electrical conductors (Thackeray 2002). The kinetics can be enhanced e.g. by doping, size-tailoring and/or "carbon nanopainting". This can yield capacities of about 160 mAh/g which is near the theoretical value of 170 mAh/g (Armand, Gauthier, Magnan, and Ravet 2002). Although controversially discussed, the olivine member LiMnPO$_4$ might be an even more promising cathode material than LiFePO$_4$ due to the higher operational voltage of 0.6–0.7 V since it offers a rather flat discharge voltage curve at 4.1 V (Padhi, Nanjundaswamy, and Goodenough 1997). However, this material particularly suffers from poor electronic and ionic conductivity and increasing research activities aim to overcome this problem, e.g., by synthesis of carbon-coated nanostructured material with stable reversible capacity up to 145 mAh/g (Wang, Buqa, Crouzet, Deghenghi, Drezen, Exnar, Kwon, Miners, Poletto, and Grätzel 2009). In this respect, it is a promising task to synthesize nanoscaled LiMPO$_4$ with various transition metal oxides, such as M = Mn, Co. The low intrinsic electronic and ionic conductivity of these materials in bulk requires particles with a large surface to volume ratio in order to reduce ionic or/and electronic diffusion lengths, and possibly optimized post-synthesis treatment like carbon coating.

2 EXPERIMENTAL

The synthesis and detailed characterization of the materials will be presented elsewhere. The post treated active material was applied on aluminium nets, pressed to plates and placed in swagelok-type cells with a Li-foil counter electrode under argon atmosphere. The electrolyte was a 1 M solution of $LiPF_6$ in 1:1 Ehtylenecarbonate and Dimethylcarbonate (Electrolyte LP30, Merck) introduced in glass microfibre seperators (Whatman, Schleicher & Schuell). Electrochemical measurements were carried out using a VMP3 Potentiostat System with EC-Lab v10 software in an incubator at 25°C. The cells were cycled in the range from 3.5 to 4.5 V, PEIS (potentioelectrochemical impedance spectroscopy) and Staircase PEIS measurements were performed in the range from 1 mHz to 200 kHz with an amplitude of 10 mV.

3 RESULTS AND DISCUSSION

Figure 1 shows Nyquist plots and cyclic voltamograms (CVs) as well of as-synthesized as of post-synthesis annealed $Li/Li^+/LiMnPO_4$ cells. The CVs show a typical behavior for $LiMnPO_4$ cathodes with reaction peaks at about 4.3 V (Mn^{2+} oxidation) and 3.95 V (Mn^{3+} reduction) vs. Li/Li^+ similar to what is found in the literature (The Nam Long Doan 2011). The post-synthesis annealed material exhibits much more pronounced cathodic and anodic peaks which imply significantly higher reaction kinetics.

For both materials, the PEIS spectra show a high frequency semicircle which according to Thomas et al. (Ref. (Thomas, Bruce, and Goodenough 1985)) may be associated with charge transfer resistances between the electrolyte and the particles, and double-layer capacitance. They exhibit a high frequency ending at a purely real value of about 10 Ω which can be

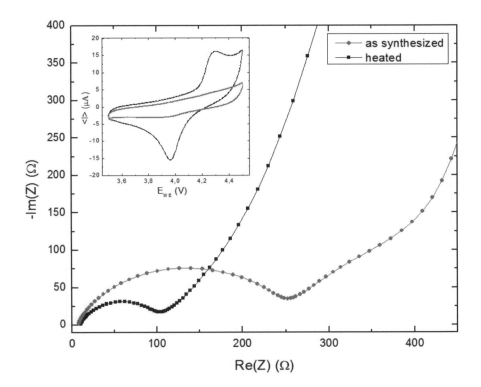

Figure 1. PEIS spectra of $Li/Li^+/LiMnPO_4$ cells equipped with as-prepared active material (red circles) and heattreated material (500°C over 4 h under low pressure argon atmosphere) (black squares). Inset: CV of both materials.

attributed to the combined electrolyte and electrode contact resistances and is dependent on the cell setup. The straight line in the low frequency area (only shown up to 0.1 Hz here) shows typical diffusion controlled behavior. The divergence from a 45° Warburg slope (semi-infinite linear diffusion) could possibly be a consequence of the finite particle size distribution (Levi and Aurbach 2004) or due to anomalous diffusion (Bisquert and Compte 2001). The comparison of as-synthesized and posttreated material shows a clear decrease of the charge transfer resistance. Furthermore, the as-synthesized material exhibits two different diffusion controlled areas, which are not observed in the spectra of the post-treated powder.

ACKNOWLEDGMENTS

Support by the BMBF via the LIB2015 alliance (independent research group, projects 03SF0340 and 03SF0397) is gratefully acknowledged.

REFERENCES

Armand, M., M. Gauthier, J.-F. Magnan, and N. Ravet (2002). Method for Synthesis of Carbon-coated Redox Materials with Controlled Size.

Bisquert, J. and A. Compte (2001). Theory of the electrochemical impedance of anomalous diffusion. *Journal of Electroanalytical Chemistry 499*(1), 112–120.

Goodenough, J.B. (2007). Cathode materials: A personal perspective. *Journal of Power Sources 174*(2), 996–1000.

Levi, M.D. and D. Aurbach (2004). Impedance of a single intercalation particle and of non-homogeneous, multilayered porous composite electrodes for Li-ion batteries. *The Journal of Physical Chemistry B 108*(31), 11693–11703.

Padhi, A.K., K.S. Nanjundaswamy, and J.B. Goodenough (1997). Phospho-olivines as Positive-Electrode Materials for Rechargeable Lithium Batteries. *Journal of the Electrochemical Society 144*, 1188.

Thackeray, M. (2002). Lithium-ion batteries: An unexpected conductor. *Nature Materials 1*(2), 81–82.

The Nam Long Doan, Izumi Taniguch, I. (2011). Cathode performance of $LiMnPO_4$/C nanocomposites prepared by a combination of spray pyrolysis and wet ball-milling followed by heat treatment. *Journal of Power Sources 196*(3), 1399–1408.

Thomas, M.G.S.R., P.G. Bruce, and J.B. Goodenough (1985). AC Impedance Analysis of Polycrystalline Insertion Electrodes: Application to LiCoO. *Journal of the Electrochemical Society 132*, 1521.

Wang, D., H. Buqa, M. Crouzet, G. Deghenghi, T. Drezen, I. Exnar, N.H. Kwon, J.H. Miners, L. Poletto, and M. Grätzel (2009). High-performance, nano-structured $LiMnPO4$ synthesized via a polyol method. *Journal of Power Sources 189*(1), 624–628.

Lecture Notes on Impedance Spectroscopy, Volume 3 – Kanoun (ed)
© 2012 Taylor & Francis Group, London, ISBN 978-0-415-64430-3

Influence of nanoscaling on the electrochemical properties of LiCoO$_2$ and TiO$_2$

Carsten Jähne & Rüdiger Klingeler
Kirchhoff Institute for Physics, University of Heidelberg, Heidelberg, Germany

Galina S. Zakharova
Institute of Solid State Chemistry, Ural Division, Russian Academy of Sciences, Yekaterinburg, Russia

Andreia I. Popa, Christine Täschner, Albrecht Leonhardt & Bernd Büchner
Institute for Solid State Research, Leibniz Institute for Solid State and Materials Research IFW Dresden, Dresden, Germany

ABSTRACT: Synthesis of nanostructured LiCoO$_2$ and TiO$_2$ via microwave assisted and conventional hydrothermal methods, respectively, is presented. The physical properties of the resulting material are characterised via X-ray powder diffraction and scanning electron microscopy. Cyclic voltammetry and Galvanostatic Intermittent Titration Technique are used to study differences of the electrochemical properties compared to bulk material.

Keywords: Hydrothermal reaction, cyclic voltammetry, Galvanostatic Intermittent Titration Technique

1 INTRODUCTION

In the recent years transition metal oxide nanomaterials have received a lot of attention because of their wide field of potential applications like sensors, solar cells, and battery materials (Umar and Hahn 2009). This interest is caused by fundamentally novel properties appearing upon size reduction which already have originated novel applications for long-known materials. One of the most promising fields of application concerns the usage of nanoscaled electrode materials in lithium-ion batteries. Here, high surface area and reduced ionic or/and electronic diffusion lengths as well as less tendency to structural phase transitions compared to the bulk form render nanoscaled materials beneficial for electrochemical energy storage (Wang, Li, He, Hosono, and Zhou 2010, Bruce, Scrosati, and Tarascon 2008).

Here, we address two materials which bulk counterparts are used in commercial lithium ion batteries. One is TiO$_2$ which is a excellent anode material if high safety, good capacity retention, low self discharge, chemical stability and negligible toxicity are demanded (Armstrong, Armstrong, Canales, and Bruce 2006, Ortiz, Hanzu, Djenizian, Lavela, Tirado, and Knauth 2009). The layer structure promotes intercalation of Li ions (Dambournet, Belharouak, and Amine 2009). The second material addressed here is LiCoO$_2$ which is one of the most widely used cathode material. It features layered structure but exhibits instability problems upon deep discharge as well as high costs and toxicity (Whittingham 2004).

2 EXPERIMENTAL

Detailed description of the synthesis procedures as well as thorough characterisation of the materials will be described elsewhere. For both materials, hydrothermal synthesis procedures have been applied. While a conventional hydrothermal technique was used for TiO_2, a novel microwave-assisted approach with precursors according to Jo et al. (Jo, Hong, Choo, and Cho 2009) has been employed for $LiCoO_2$. In order to optimize the material and to correlate morphology, size, synthesis parameters and electrochemical performance, we studied several series of samples.

Electrochmical studies by means of cyclic voltammetry and Galvanostatic Intermittent Titration Technique (GITT) have been done in two-electrode swagelok-type cells. The resulting active material was pressed on aluminium mesh and mounted with a Li-foil counter electrode under argon atmosphere. The used electrolyte was a 1 M solution of $LiPF_6$ in 1:1 Ehtylenecarbonate and Dimethylcarbonate (Electrolyte LP30, Merck), the separator was a Whatman glass microfiber. Electrochemical measurements were carried out by means of a VMP3 potentiostat system. Both materials were cycled between different potential limits according to their reported redox characteristic used in Lithium-Ion batteries.

3 RESULTS AND DISCUSSION

TiO_2 crystallises in a tube-like morphology with outer diameter of 10–15 nm and lengths of several hundred nanometres. The electrochemical processes upon cycling are visible in cyclic voltammograms (CVs) presented in Fig. 1. Beside good reversibility of the main reaction peaks attributed to the redox couple at 1.7 V/1.99 V minor differences are found between first and consecutive cycles. The GITT analyses (not shown) show excellent cyclability with a coulombic efficiency of 99% and a specific capacity after 50 cycles of 265 mAh/g which is significant higher than the bulk capacity of 170 mAh/g reported by Dambournet et al. (Armstrong, Armstrong, Canales, and Bruce 2006).

$LiCoO_2$ crystallises in a plate-like morphology with typical edge lengths of 20 nm. Comparing the first and second cycle of the CVs of $LiCoO_2$ shown in Fig. 2, a quasi-reversible character is found. This behaviour is not found for bulk material. The quasi-reversibility is accompanied by an additional oxidation peak occurring around 3.75 V.

According to Santiago et al (Santiago, Andrade, Paiva-Santos, and Bulhoes 2003) this can be attributed to a low temperature spinel phase. The GITT spectra (not shown here) feature a vanishing plateau region and exhibit a slightly lower specific capacity compared to bulk material, similar to what is reported by Okubo et al (Okubo, Hosono, Kudo, Zhou, and Honma 2009).

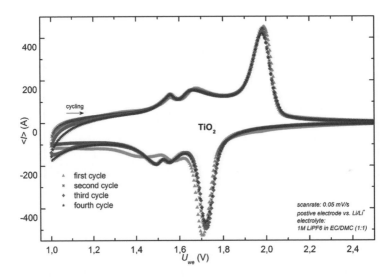

Figure 1.   CV spectra of nano structured TiO_2.

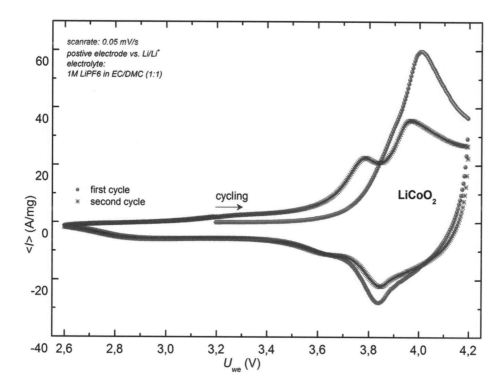

Figure 2. CV spectra of nano structured LiCoO$_2$.

In summary, self-organized transition metal oxide nanostructures were successfully synthesised by means of hydrothermal methods. In particular, the microwave-assisted method provides a fast and reliable technique to produce reasonable amounts of material. The nano-sized morphology of the materials can influence the reaction behaviour so that higher capacities can be reached (TiO$_2$) than in the bulk or metastable phases can be stabilised (LiCoO$_2$).

ACKNOWLEDGMENTS

Support by the BMBF via the LIB2015 alliance (independent research group, projects 03SF0340 and 03SF0397) is gratefully acknowledged.

REFERENCES

Armstrong, G., A.R. Armstrong, J. Canales, and P.G. Bruce (2006). TiO$_2$(B) Nanotubes as Negative Electrodes for Rechargeable Lithium Batteries. *Electrochemical and Solid-State Letters 9*, A139.

Bruce, P.G., B. Scrosati, and J.M. Tarascon (2008). Nanomaterials for rechargeable lithium batteries. *Angewandte Chemie International Edition 47*(16), 2930–2946.

Dambournet, D., I. Belharouak, and K. Amine (2009). Tailored Preparation Methods of TiO$_2$ Anatase, Rutile, Brookite: Mechanism of Formation and Electrochemical Properties. *Chemistry of Materials 22*(3), 1173–1179.

Jo, M., Y.S. Hong, J. Choo, and J. Cho (2009). Effect of LiCoO Cathode Nanoparticle Size on High Rate Performance for Li-Ion Batteries. *Journal of the Electrochemical Society 156*, A430.

Okubo, M., E. Hosono, T. Kudo, H.S. Zhou, and I. Honma (2009). Size effect on electrochemical property of nanocrystalline LiCoO$_2$ synthesized from rapid thermal annealing method. *Solid State Ionics 180*(6–8), 612–615.

Ortiz, G.F., I. Hanzu, T. Djenizian, P. Lavela, J.L. Tirado, and P. Knauth (2009). Alternative li-ion battery electrode based on self-organized titania nanotubes. *Chemistry of Materials 21*(1), 63–67.

Santiago, E.I., A.V.C. Andrade, C.O. Paiva-Santos, and L.O.S. Bulhoes (2003). Structural and electrochemical properties of $LiCoO_2$ prepared by combustion synthesis. *Solid State Ionics 158*(1), 91–102.

Umar, A. and Y. Hahn (2009). *Metal Oxide Nanostructures and Their Applications*. Amer Scientific Publishers.

Wang, Y., H. Li, P. He, E. Hosono, and H. Zhou (2010). Nano active materials for lithium-ion batteries. *Nanoscale 2*(8), 1294–1305.

Whittingham, M.S. (2004). Lithium batteries and cathode materials. *Chemical Reviews 104*(10), 4271–4302.

Lecture Notes on Impedance Spectroscopy, Volume 3 – Kanoun (ed)
© *2012 Taylor & Francis Group, London, ISBN 978-0-415-64430-3*

Characterization of CVD/ALD layers with impedance sensors

Konrad Hasche, Michael Hintz, Stefan Völlmeke, Arndt Steinke & Ingo Tobehn
CiS Forschungsinstitut für Mikrosensorik und Photovoltaik GmbH, Erfurt, Germany

ABSTRACT: At CIS, we develop a non-destructive and, ultimately, non-contact method to measure CVD and ALD layers. These layers are well established in the manufacturing processes of the semiconductor industry for example. We want to monitor the process and scan their thicknesses and relative permittivities. For this we use modified Interdigital structures with a few microns distance. CiS produces such structures for a long time—eg to produce humidity sensors.

1   BASIC PRINCIPLES OF THE MEASUREMENT

For the measurements, no arrangement has used in form of a plate capacitor, in which you must place the material to be measured between the plates (electrodes), but a planar interdigital structure in micro system technology. This planar electrode structure can be positioned parallel to the surface of the medium being measured, so that the electric field detected is influenced. The frequency of the applied field varies from 1 kHz up to several MHz and some appropriate meaningful metrics, such as the capacity or the phase angle or the complex impedance are displayed in an impedance spectrum. In this context it is intended to modify the arrangement so that measurements are possible at higher frequencies. Depending on the thickness of the deposited layer penetrated by the electric field the quantity detected by the sensor is composed of two components. If the coating is completely missing a measurement signal influenced only by the relative permittivity of the substrate is measured. At the other hand, the carrier material can be neglected if the coating is thick enough.

In this context, using ANSYS®, numerous simulations were performed. The properties of the substrate, the coatings and the geometry of the interdigital structures were varied. Generally, there exists functional dependencies such as capacity (thickness) as continous functions, so that you can specify them using a sufficient number of basic values that can be determined numerically. In the Figures 1 and 2 for the simulation silicon dioxide as a carrier material with $\varepsilon_r = 4.0$ was assumed and a coating with $\varepsilon_r = 80$, which is equivalent to TiO_2. Initially it was assumed that the sensor rests directly on the wafer to be measured. The fact that in practice always a small air gap is present appears for the practice not ultimately relevant, because the process is already feasible only by differential measurements.

In future sensor assemblies should already be constructed by an additional integrated optical sensor, so that the method works without contact. The probe has always a slight distance from the object in this case. Currently it is guaranteed through an integrated pressure sensor that acts in this case as a touch sensor, that the sensor rests on the surface.

2   EXPERIMENTAL SETUP

As a fundamental element, a new Mesa—structure in which the interdigital structure is located on a plateau about 200 μm is elevated to the rest of the chip as shown in figure 3

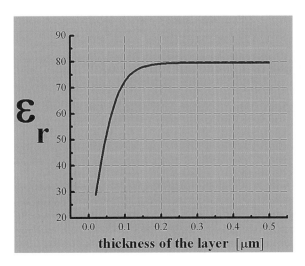

Figure 1.   Relative permittivity as a function of layer thickness with an interdigital structure electrode width of 50 nm and 50 nm electrode spacing.

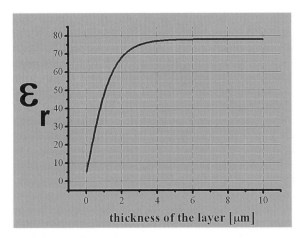

Figure 2.   Relative permittivity as a function of layer thickness with an interdigital structure electrode width of 2.5 μm and 1.5 μm electrode spacing.

(Völlmeke, Preuß, and Steinke 2010). This ensures that the sensitive plane of the sensor also can really rest on the specimen. The metallization (2 μm to 20 μm track spacing depending on sensor) consists of $MoSi_2$ and is covered by a planarized passivation of SiO_2 and Si_xN_y.

The main components of the structure are shown in figure 4. The sensor carrier is mounted on the z-axis of a CNC assembly and can be lowered and attached to the wafer to be measured. At the bottom of the carrier, the interdigital structure is placed and a modified pressure sensor from CiS production (CDW 02) (Semmler 2011) as a control component, which was sealed with an elastic material is installed. In this way, the wafers are scanned to obtain a better statistical relevance. The chuck for holding the wafer 4 inch is made of PEEK (polyetheretherketon).

The wafers used have coatings of Al_2O_3, YSZ, TiO_2 or $SrTiO_3$ with thicknesses of 50, 100, 200 and 500 nm, and relative permittivities between 10 and 150. The coatings were assembled by the Fraunhofer IKTS in Dresden (Endler and Rose 2007).

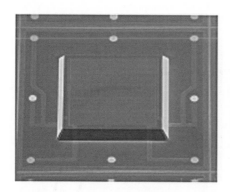
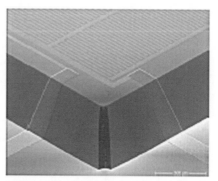

Figure 3. Planar interdigital mesa structure (Völlmeke, Preuß, and Steinke 2010).

Figure 4. Experimental set-up, description of components see text.

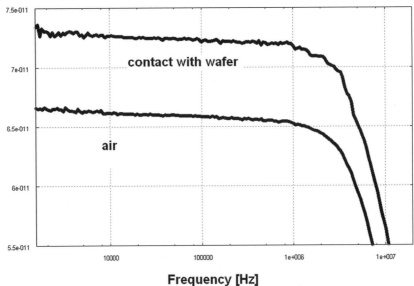

Capacity Cp [F]

Frequency [Hz]

Figure 5. Capacity C_p as a function of frequency without wafer (bottom) and with a coated wafer (47 nm $SrTiO_3$ with $\varepsilon_r \approx 150$).

3 PRELIMINARY RESULTS

First measurements are shown in Figure 5, which were carried out with an interdigital structure with 10 μm electrode spacing and electrode width of 7 μm. To receive the lower curve, the sensor head was about 3 cm above the wafer. Making the effective dielectric in this case was only air. For the recording of the upper curve the sensor was placed directly on the silicon wafer which was coated in this case with 47 nm $SrTiO_3$.

For an optimization and better elaboration of the measurement effect, depending on the thickness of the coating and its permittivity further modifications to the test setup and simulation are necessary.

REFERENCES

Endler, I. and M. Rose (2007). Jahresbericht 2007: Nanoskalige ald-schichten mit hoher dielektrizität-skonstante für speicherkondensatoren. Technical report, Fraunhofer-Institut für Keramische Technologien und Systeme.

Semmler, K. (2011). Geschäftsfeld mems. Technical report, CiS Forschungsintitut für Mikrosensorik und Photovoltaik GmbH.

Völlmeke, S., K. Preuß, and A. Steinke (2010). Technologiekompatible 3d-strukturierung zur herstellung integrierter mikrosysteme. In *Technologien und Werkstoffe der Mikrosystem- und Nanotechnik—2. GMM-Workshop*. VDE VERLAG GmbH.

Lecture Notes on Impedance Spectroscopy, Volume 3 – Kanoun (ed)
© 2012 Taylor & Francis Group, London, ISBN 978-0-415-64430-3

Author index